A Hiking Guide to the Geology
of the Wasatch Mountains

A Hiking Guide to the Geology of the Wasatch Mountains

Mill Creek and Neffs Canyons,
Mount Olympus, Big and Little
Cottonwood and Bells Canyons

by

William T. Parry

THE UNIVERSITY OF UTAH PRESS
Salt Lake City

16 15 14 13 12   5 4 3 2

 The Defiance House Man colophon is a registered trademark of the University of Utah Press. It is based upon a four-foot-tall, Ancient Puebloan pictograph (late PIII) near Glen Canyon, Utah.

LIBRARY OF CONGRESS CATALOGING-IN-PUBLICATION DATA

Parry, William T.
 A hiking guide to the geology of the Wasatch mountains : a geological guide to the hiking trails in Mill Creek, Big Cottonwood, and Little Cottonwood Canyons / by William T. Parry.
      p. cm.
 Includes bibliographical references and index.
 ISBN-13: 978-0-87480-839-1 (pbk. : alk. paper)
 1. Geology—Wasatch Range (Utah and Idaho) 2. Geology, Stratigraphic.
3. Paleogeography—Wasatch Range (Utah and Idaho) 4. Wasatch Range
(Utah and Idaho)—Guidebooks. I. Title.
 QE652.55.W37P37 2005
 557.92'2—dc22   2005016948

Interior printed on recycled paper with 60% post-consumer content.

This book is dedicated to Gale Dick, Alexis Kelner, and all of the volunteers of the Citizens Committee to Save Our Canyons.

Contents

Maps

Acknowledgments

The author gratefully acknowledges the numerous discussions of Wasatch Mountain geology with faculty colleagues and students over many years. The careful reviews and proofreading of Margie Chan, Bill Case, Gayle Parry, and the University of Utah Press editors substantially improved the text and figures. The helpful suggestions of Alexis Kelner are also gratefully acknowledged.

Introduction

The Wasatch Mountains rise spectacularly from the relatively flat floor of the valley in which Salt Lake City is located. Thirteen peaks over 11,000 feet in elevation and another nineteen peaks more than 10,000 feet in elevation make the rugged topography of the Wasatch Mountains one of the largest features in the relief of Utah. Easy access into the mountain range is provided by paved highways and numerous hiking trails. Many visitors to the canyons and trails have enjoyed the rugged scenery for more than a century. This guide is meant to enrich the experience of visitors and hikers by providing the background to understand the geological history and development of the Wasatch Mountains.

The Central Wasatch Mountains provide a unique opportunity to observe the exposed record of development of a continental margin. A record of the continually varying positions of the continent is preserved in the rocks. The rock record includes the assembly and later breakup of at least two supercontinents, several major mountain-building events, and glacial erosion and deposition that began more than 700 million years ago. Igneous intrusions and formation of rich silver ores are among the many geological features that can be observed along the hiking trails. Faulting and earthquake activity continue today. The descriptive trail guide to the main hiking trails is presented in this book to assist the hiker and naturalist in observing the diverse geological history of the area.

This trail guide presents the record of rock accumulation in ascending order from oldest to youngest, the record of mountain-building events from oldest to youngest, the effects of glaciation, and the development of the present topography. The guide then describes the geology of Mill Creek, Big Cottonwood, and Little

Cottonwood Canyons, followed by a description of the specific geology and locations of major hiking trails with trailheads in each of these canyons. Although some potential hiking routes are not included, the experienced hiker could easily piece together many additional routes from the trail guides provided. A summary of trail length, elevation gain, relative difficulty, and major geological features of each trail is shown in Table 1.

Trail	One Way Length	Elevation Gain	Trail Difficulty	Geological Features
Mill Creek Canyon Trails				
Grandeur Peak from the West	1.75 miles	3100 feet	Difficult	Thaynes Formation, Ankareh Formation
Church Fork to Grandeur Peak	2.75 miles	2619 feet	Difficult	Thaynes Formation, Ankareh Formation, Nugget Sandstone, small fault, spring tufa
Thaynes Canyon to Neffs Canyon Divide	2.25 miles	2960 feet	Difficult	Park City Formation, Weber Quartzite, Round Valley Limestone, Dough-nut Formation, Humbug Formation, Mount Raymond thrust fault
Salt Lake Overlook	1.75 miles	1250 feet	Easy	Weber Quartzite, Park City Formation, small thrust fault
Pipeline—Rattle-snake Gulch to Church Fork	2.1 miles	700 feet	Easy	Park City Formation, Woodside Shale
Pipeline— Church Fork to Birch Hollow	1.5 miles	320 feet	Easy	Park City Formation, Woodside Shale
Pipeline— Birch Hollow to Elbow Fork	2.2 miles	620 feet	Easy	Park City Formation, Woodside Shale, Thaynes Formation, fault at mile 1.5
Elbow Fork to Mount Aire	1.8 miles	1991 feet	Moderate	Thaynes Formation, Ankareh Formation, Nugget Sandstone
Elbow Fork to Lambs Canyon	1.5 miles	1490 feet	Easy	Thaynes Formation, Ankareh Formation
Porter Fork to Big Cottonwood Divide	3.25 miles	3660 feet	Difficult	Park City Formation, Weber Quartzite, Round Valley Limestone, Doughnut Formation, Mount Raymond thrust (2 strands), Humbug Formation, Deseret Formation

Table 1. *The Trails*

Trail	One Way Length	Elevation Gain	Trail Difficulty	Geological Features
Bowman Fork to Baker Pass	3.75 miles	3080 feet	Difficult	Park City Formation, Weber Quartzite, Round Valley Limestone, Humbug Formation, Doughnut Formation, Mount Raymond thrust fault
Alexander Basin to Gobblers Knob	2.6 miles	3086 feet	Difficult	glacial moraine, Weber Quartzite, Round Valley Limestone
Big Water to Dog Lake Junction	2.5 miles	1200 feet	Easy	Park City Formation, Nugget Sandstone, Mount Raymond thrust fault, Silver Fork fault
Little Water to Dog Lake Junction	2 miles	1200 feet	Easy	Weber Quartzite, Park City Formation, Thaynes Formation, Ankareh Formation, glacial moraine, Silver Fork fault, Mount Raymond thrust fault
Upper Mill Creek to Wasatch Crest	3 miles	1900 feet	Moderate	Park City Formation, Woodside Shale, Thaynes Formation, glacial moraine, Mount Raymond thrust fault (hidden)
Mountain Front Trails				
Neffs Canyon to Thaynes Canyon Pass	3.5 miles	3190 feet	Difficult	Humbug Formation, Gardison Formation, Deseret Formation, Fitchville Formation, Tintic Quartzite, Mount Raymond thrust fault
Mount Olympus	3 miles	4050 feet	Very Difficult	Mutual Formation, Big Cottonwood Formation, Wasatch fault
Big Cottonwood Canyon Trails				
Mill B North Fork to Mill Creek Divide	3.25 miles	2960 feet	Difficult	Mineral Fork Tillite, Big Cottonwood Formation, Alta thrust fault, Tintic Quartzite to Doughnut Formation
Mill B South Fork to Lake Blanche	2.75 miles	2720 feet	Moderate	glacial moraine, Big Cottonwood Formation, Mineral Fork Tillite boulders, glacial grooves

Table 1. *Continued*

Trail	One Way Length	Elevation Gain	Trail Difficulty	Geological Features
Broads Fork to end of the trail	2 miles	2200 feet	Moderate	glacial moraine, Big Cottonwood Formation
Mineral Fork to Wasatch mine	3 miles	1940 feet	Moderate	Big Cottonwood Formation, Mineral Fork Tillite, glacial moraine, Silver King fissure
Butler Fork to Baker Pass Divide	3.5 miles	2050 feet	Difficult	Weber Quartzite, Park City Formation, Woodside Shale, Weber Quartzite, Mount Raymond thrust fault
Cardiff Fork to Cardiff Pass	3.25 miles	2720 feet	Difficult	Deseret and Gardison Limestone, Tintic Quartzite, Mineral Fork Tillite, Alta thrust fault, Superior fault
Dog Lake	2 miles	1520 feet	Easy	Glacial moraine, Gardison Limestone to Park City Formation, Woodside Shale, Thaynes Formation, Silver Fork fault, Mount Raymond thrust fault
Desolation Lake	3.5 miles	1890 feet	Moderate	glacial moraine, Gardison Limestone to Park City Formation, Woodside Shale, Thaynes Formation, Ankareh Formation, Silver Fork fault
Days Fork to Silver Fork Pass	3 miles	2570 feet	Difficult	glacial moraine, Tintic Quartzite, Deseret Limestone, Humbug Formation, Gardison Limestone, Silver Fork fault
Silver Fork to Prince of Wales mine	3 miles	1500 feet	Moderate	glacial moraine, Silver Fork fault, Alta thrust fault, Deseret and Gardison Limestones, Fitchville Formation
Little Willow to Willow Lake	0.75 miles	640 feet	Easy	glacial moraine, glacial lake
Brighton to Clayton Peak by way of Snake Creek Pass	2.5 miles	1960 feet	Moderate	glacial moraine, igneous intrusive of the Clayton Peak stock

Table 1. *Continued*

Trail	One Way Length	Elevation Gain	Trail Difficulty	Geological Features
Brighton to Lake Mary	1 mile	760 feet	Easy	igneous intrusive of the Alta stock, glacial moraine
Brighton to Lake Catherine	2 miles	1200 feet	Easy	igneous intrusive of the Alta stock, glacial moraine, Gardison Limestone
Brighton to Twin Lakes	1.25 miles	710 feet	Easy	Deseret and Gardison Limestone, igneous intrusive of the Alta stock, glacial moraine
Mountain Front Trail				
Bells Canyon to upper Bells reservoir	3.75 miles	4100 feet	Very Difficult	glacial moraine, igneous intrusive of the Little Cottonwood stock, Wasatch fault
Little Cottonwood Canyon Trails				
White Pine Canyon to White Pine Lake	4.5 miles	2460 feet	Difficult	glacial moraine, igneous intrusive of the Little Cottonwood stock
Red Pine Canyon to Red Pine Lake	3 miles	1940 feet	Difficult	glacial moraine, igneous intrusive of the Little Cottonwood stock
Cecret Lake	0.75 miles	420 feet	Easy	glacial moraine, Tintic Quartzite, igneous dikes, Mineral Fork Tillite, sulfide mineralization on fractures
Catherine Pass	1 mile	800 feet	Easy	Alta stock igneous intrusive, Tintic Quartzite, Fitchville Formation, Gardison and Deseret Limestones, Maxfield Limestone, Alta thrust fault, contact metamorphism
Grizzly Gulch to Twin Lakes Pass	1.75 miles	1353 feet	Easy	Ophir Shale, Maxfield Limestone, Fitchville Formation (ore bed), Gardison Limestone, Deseret Limestone, igneous intrusive of the Alta stock, Silver Fork fault, Alta thrust fault, contact metamorphism

Table 1. *Continued*

Geology of the
Wasatch Mountains

Descriptions of the geology and geologic history of Utah that include the Wasatch Mountains can be found in Stokes (1986) and Hintze (1988). Granger and others (1952) described the geology of these mountains, but their description preceded our present understanding of plate tectonics. The Wasatch Mountains occur at a junction of Basin and Range and Colorado Plateau where the geology changes abruptly from the folded and thrust-faulted rocks of the Wasatch to the relatively undeformed rocks of the Colorado Plateau. The topography that we see on the skyline looking eastward from Salt Lake City is jagged and rough, consisting of high peaks separating stream- and glacier-cut canyons as shown in Figure 1. The rugged mountains consist of folded and faulted Mesozoic, Paleozoic, and Precambrian rocks as shown in Figure 2. The mountains are steep with bedrock cliffs, narrow gorges, and jagged summit ridges. Alluvial fans and Lake Bonneville deltas form the boundary between mountains and the flat lake plain that is the valley floor. Alluvial fans are faulted and merge with deltas below the level of ancient Lake Bonneville. The canyons extend northeastward from the mountain front and follow the sedimentary rock layers that were folded into several northeast-trending folds. The high elevation of the mountains formed a relatively short time ago, and the sculpting of the high elevation by streams and glaciers is even more youthful. Both the geological uplift and the erosion are still in progress. The Wasatch Mountains are youthful, but this apparent youth masks older mountain-building events, mountains whose jagged outlines are

Figure 1. *View of the Wasatch Mountains and Salt Lake Valley from Neffs Canyon on the north to Lone Peak on the south looking southeast.*

long gone, leaving only the mountain roots as part of the building blocks of the present-day Wasatch Mountains.

A series of compressional mountain-building events progressed from California across Nevada and Utah to Colorado from Late Devonian to Early Cenozoic time. The earth's crust from California's Sierra Nevada Mountains to Colorado's Rocky Mountains was thickened as a result. By the end of the Cretaceous, a plateau 2.5 miles high was present in western North America. Since then, the thickened crust in Utah and Nevada has been broken, extended, and thinned so that the high plateau is no longer present. The high elevation of the Wasatch Mountains is not supported by thickened crust alone, however. The elevated topography is produced by heated, low-density, buoyant mantle.

Many of the features of the Central Wasatch Mountains are explained in terms of the theory of plate tectonics. According to this theory, the outer, most rigid layers of the earth are divided into a series of plates. The boundaries between plates may be in

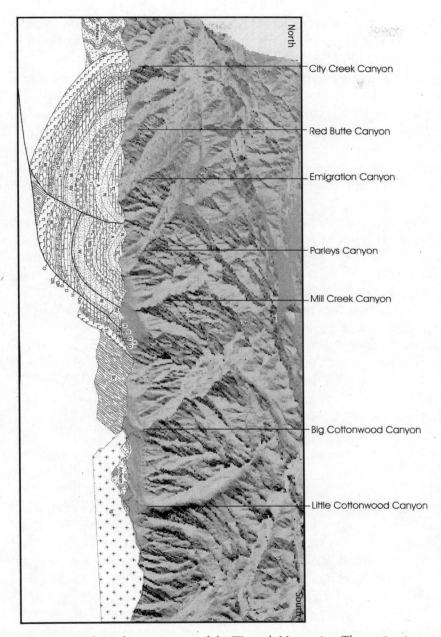

Figure 2. *Geological cross section of the Wasatch Mountains. The section is a north-south cross section looking eastward after Bryant (1990). Formation patterns and symbols are shown in Figure 5.*

9

the ocean basins, on margins of continents, or within continents. The plates are in motion with respect to one another. The motion is driven by heat from the interior of the earth. When some plates collide with their neighbors, the collision zone forms mountain ranges. As other plates diverge from their neighbors, the divergence produces rift valleys. Some plate margins are neither convergent nor divergent but slide past one another. The constant motion of the plates has resulted in the formation and destruction of ocean basins, the formation of large assemblages of continental fragments into supercontinents, and the destruction of supercontinents.

At least two supercontinents are known. The oldest known supercontinent, though probably not the first one to form, is called Rodinia. The continental fragments were assembled into Rodinia by about 1,100 million years ago. The approximate configuration of Rodinia is shown in Figure 3. The continental fragments that comprised Rodinia included fragments of most of the continents known on the earth today. Laurentia is the ancient continent incorporating what is now North America that rifted from Rodinia. Rodinia broke apart beginning about 750 million years ago, and the fragment that was to become North America had rifted off by about 550 million years ago. Fragments of Rodinia moved about the globe as ocean basins formed and were destroyed until collisions between separate fragments began the assembly of a new supercontinent known as Pangea that straddled the equator (Dalziel 1995). Pangea was assembled in stages. Gondwana was a supercontinent that formed 750 to 550 million years ago, consisting of Africa, South America, Antarctica, India, and a few other fragments that later became part of Pangea. By about 260 million years ago, the assembly of Pangea was complete. The configuration of Pangea is shown in Figure 4. This supercontinent also broke apart beginning about 200 million years ago when the Atlantic Ocean intervened between Africa and South America.

The consequence of motion of this portion of the North American continent was repeated collisions with oceanic or continental lithosphere that produced episodic mountain building.

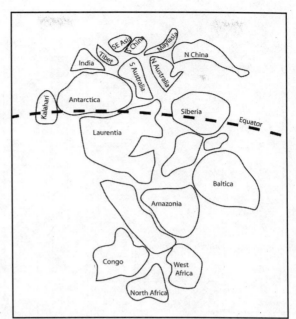

Figure 3. *Approximate configuration of Rodinia at 1,000 million years ago, modified from Condie (1997). The continental fragments that made up the supercontinent are identified with present-day geographical names.*

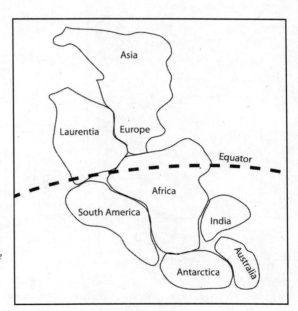

Figure 4. *Approximate configuration of Pangea about 200 million years ago, modified from Condie (1997). The continental fragments that made up the supercontinent are identified with present-day geographical names.*

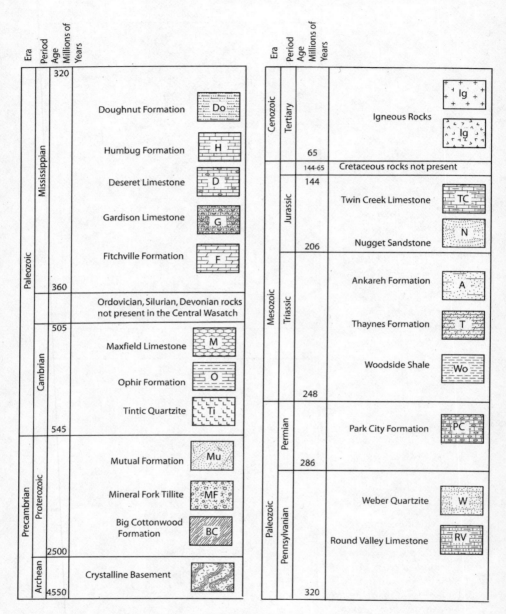

Figure 5. *List of the sedimentary stratigraphy and igneous rocks exposed in the Wasatch Mountains. The list includes the geologic age and absolute age of each interval of time. Each rock unit is shown with the map pattern used to indicate outcrop areas on the geologic maps.*

A further consequence of the motion was exposure of the continent to different climate regimes that also left a record in the rocks.

The Wasatch Mountains from Nephi to the Idaho border expose older sedimentary, igneous, and metamorphic rocks in an erosional window through young cover. The older rocks have been exhumed and exposed by erosion. The window north and south of Salt Lake City is very narrow following the ridge crest of the Wasatch, but east of Salt Lake City the window broadens eastward toward the Uinta Mountains. This broadened window is due to exhumation of the Cottonwood uplift, an extension of the Uinta uplift to the east.

The sedimentary and igneous rocks exposed in the Wasatch, together with their ages, are shown in Figure 5. The oldest rock exposed in the area covered by this trail guide is the Early Proterozoic-age Little Willow Formation. The youngest sedimentary rock unit exposed in the area covered by this guide is the Jurassic-age Twin Creek Limestone. Thus, the rock record represents nearly 2 billion years of earth history. Younger rocks and older rocks that are exposed elsewhere in the Wasatch are not described in this guide. Each rock unit has been interpreted in terms of its significance in the geological history.

The rocks have been involved in a series of mountain-building events that have left a record in the rocks. The folding, faulting, and igneous intrusions in the Wasatch date from earlier mountain building. The youngest mountain-building event continues today.

Wasatch Mountain Rocks

Precambrian Crystalline Basement

Precambrian rocks are the oldest rocks on earth, and *basement* refers to the oldest and deepest rocks present in the Wasatch Mountains. These rocks are composed of minerals that have been formed and enlarged by the heat and pressure accompanying metamorphism. The mineral crystals are often large enough to permit identification with a simple magnifying lens, in contrast to many sedimentary rocks that are composed of finely divided minerals requiring a powerful microscope to observe. Hence, the term "crystalline basement rocks" is applied.

The oldest rock in the Wasatch Mountains, the Precambrian crystalline basement, is Archean in age, rock that is older than 2.5 billion years. The oldest rock, exposed north of Salt Lake City, is called the Farmington Complex after its occurrence in Farmington Canyon. The Farmington Complex is composed of crystalline rocks that are a part of the Wyoming continental nucleus or craton. The crystalline basement extends westward nearly all the way to Elko, Nevada. Beyond Elko, no Precambrian basement exists.

Two kinds of Precambrian rocks in Utah are exposed in the Wasatch Mountains: the crystalline basement, which is very old, and the younger, lesser metamorphosed sedimentary rocks that were deposited in Precambrian time about a billion years ago. The whole region of Utah is underlain by Precambrian crystalline basement. The younger, Precambrian sedimentary sequence lies above the Archean crystalline basement.

The younger Precambrian rocks are Proterozoic in age and are exposed in the core of the Uinta Mountains where they are called the Uinta Mountain group, and at the mouth of the Big Cottonwood Canyon where they are called the Big Cottonwood Canyon group. Another Precambrian unit lies on top of the Big Cottonwood Canyon formation called the Mineral Fork Tillite. The Mineral Fork Tillite, exposed in Mineral Fork of Big Cottonwood Canyon, is significant because it represents a Precambrian glacial deposit. A distinctive purple quartzite known as the Mutual Formation lies above the Mineral Fork Tillite. That brings us up to the big Precambrian-Cambrian unconformity, the erosion surface on top of the younger Precambrian rock. Sometimes the younger Precambrian rocks were stripped away completely and the Cambrian rocks were deposited directly on the crystalline basement.

Farmington Complex

The oldest rocks in the Wasatch Mountains are very old indeed, but their maximum age is somewhat uncertain because these rocks have been buried to great depths and intruded by igneous rocks. The processes of deep burial and igneous intrusion reset the radioactive geochronometers that are used for absolute rock dating. The rocks are complexly deformed and highly metamorphosed. As a consequence they have been heated and reheated to temperatures near their melting point. Determining the age of rocks that have been repeatedly reheated is a difficult problem with the traditional radiometric dating methods in use today.

Precambrian crystalline basement in Utah consists of at least three separate crustal fragments sutured together. The approximate configuration and age of the fragments and suturing events are shown in Figure 6. First and oldest is the Farmington Complex, which is the southern part of the Wyoming continental fragment and is greater than 2,500 million years old.

Precambrian rocks north of Salt Lake City belong to this Wyoming continental fragment. These rocks began as erosional debris deposited on oceanic crust on the sea floor perhaps as long

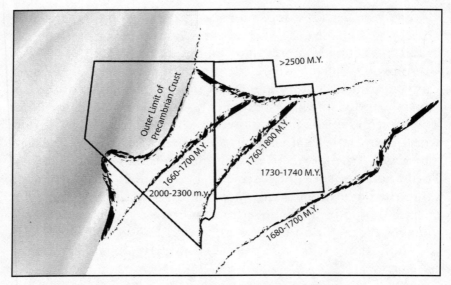

Figure 6. *The Precambrian-age crust that makes up Utah and surrounding areas.*

as 3,600 million years ago. The sediment accumulation was meta-morphosed 2,600 million years ago, intruded by granites 1,800 million years ago, and metamorphosed again between 1,650 and 1,600 million years ago. The accumulation of sediments, metamorphism, and igneous activity has produced a complex sequence of granite, granite gneiss, migmatite, schist, quartzite, amphibolite, and pegmatite of the Farmington Complex (Bryant 1988).

Precambrian rocks south of Salt Lake City are generally less than 1,800 million years old. Exposures of rocks at Santaquin south of Salt Lake represent a volcanic arc accreted to the Wyoming craton between 1,650 and 1,700 million years ago.

Comparison of the lithology and ages of ancient rocks in the Wasatch Mountains with those of the Wind River Range in Wyoming provides information on how and when the ancient continental nucleus formed. A great suture zone between Utah and Wyoming is the zone where the southern basement rocks were attached to the Wyoming craton.

The Little Willow Series is exposed in the core of the Big Cottonwood uplift just north of the mouth of Little Cottonwood Canyon, and is the oldest rock exposed in the Central Wasatch Mountains east of Salt Lake City. The Little Willow Series consists of quartzites, banded or gneissic quartzites, quartz-mica schists, and stretched pebble schists. Basic or mafic igneous rocks that are now altered to amphibolite and chlorite amphibole schists intrude them. The true age of these rocks is obscured by the intense metamorphism, including the latest metamorphic event caused by the nearby Little Cottonwood stock. The best available estimate of the age is probably early Proterozoic.

These rocks show a striking contrast in metamorphic grade and deformation with the overlying Big Cottonwood Formation. They are strongly folded and intensely metamorphosed. The overlying Big Cottonwood series is only mildly metamorphosed and gently folded. Therefore, the mountain-building event that produced the folding and metamorphism predated the deposition of the Big Cottonwood Formation.

The Farmington Complex and Little Willow Series predate the formation of the North American continent. The younger metamorphism accompanied mountain-building events associated with accretion of a continent.

Precambrian Sedimentary Rocks

Younger Precambrian sedimentary rocks that show a remarkable contrast in age, deformation, and metamorphism overlie Precambrian crystalline basement rocks. Three units of younger Precambrian rocks are present: the Big Cottonwood Formation, the Mineral Fork Tillite, and the Mutual Formation.

The Big Cottonwood Formation consists of about 16,000 feet of shale and quartzite. The entire thick sequence contains ripple marks, mud cracks, and other evidence of deposition in shallow water.

The Big Cottonwood Formation was deposited in an estuary leading eastward toward the Uinta Mountains from the open sea to the west. Kent Condie and coworkers (2001) have identified

Figure 7. *The Precambrian-age, Big Cottonwood Formation exposed in the mouth of Bells Canyon. View looking northward from the Bells Canyon trail.*

the age of the source rocks and characteristics of the source area. The source of sediment was Paleoproterozoic-age rocks that were exposed in a northwest-trending uplift in northwest Utah and highly weathered in a tropical to subtropical climate near the equator.

The shallow water deposits show a rhythmically alternating sequence of fine silt and coarser sand that was produced by ancient tides. Charles P. Sonett from the University of Arizona, Marjorie A. Chan from the University of Utah, and coworkers (1996, 1998) have found tidal rhythmites in the Big Cottonwood Formation that indicate the rate of retreat of the lunar orbit and a change in the length of the terrestrial day 900 million years ago. Measurement of the daily, monthly, and annual tidal cycles in the rhythmically layered rocks show that the day was 18 hours long and there were 481 days per year.

The Mineral Fork Tillite

Hiking in Mineral Fork of Big Cottonwood Canyon, or in Mill B North Fork or on Mount Superior, an unusual black rock containing rather large cobbles of quartzite and other rock types can be seen. This rock, the Mineral Fork Tillite, records an unusual episode in the history of the earth's climate. A spectacular exposure of the tillite over 1,000 feet thick occurs near Superior Peak. This unusual rock consists of large and small boulders in a fine-grained, dark-colored matrix and can appear like a chunk of asphaltic concrete from a highway. The boulders are mostly purplish and white quartzite similar to the quartzite in the underlying Big Cottonwood Formation. A small proportion of the boulders are light-gray limestone, calcareous and siliceous rocks, and a few granitic rocks. The source of the limestones and granites is unknown.

Microorganisms are extremely abundant in the black, fine-grained matrix of the Mineral Fork Tillite. Organic carbon from these organisms gives the rock its black color. A. H. Knoll and coworkers (1981) described these organisms in Big Cottonwood Canyon. There are as many as 1.6 million individual organisms per cubic inch of rock, a rock volume not much larger than the last segment of your little finger. The most common is a simple, unornamented unicell one–ten-thousandth to one-thousandth of an inch in size. The most distinctive organic entity is *Bavlinella faveolata*. Each individual consists of several dozen to more than 100 cells about one-thousandth of an inch in size packed together in raspberry-like multispheres. All of the forms represent different stages in the life cycle of a single category of planktonic organism that completely dominated the Mineral Fork. The organism cannot be assigned to any living category and therefore must be classed as an acritarch. Acritarchs are spherical, organic-walled microfossils. They may be dinoflagellate hystrichospheres (cyst stage) or other forms of algae that also form cysts and could include eukaryotic algae. The fossils suggest a fjord ecosystem that was environmentally stressed due to low temperature, salinity variations, and calving icebergs.

Figure 8. *The Mineral Fork Tillite exposed near Split Rock Bay on Antelope Island.*

Joseph L. Kirschvink and coworkers (2000) have shown that several intervals of intense glaciation occurred during the Precambrian. Nicholas Christie-Blick (1983) described the Mineral Fork as a well-preserved ancient example of sediment that accumulated in glacial marine environment and beneath an ice sheet at times grounded below sea level. Evidence for a glacial marine origin include the absence of clastic varves, sedimentary structures, layering pattern, dropstones, rhythmites, and laminates that indicate subaqueous deposition. Clastic varves usually form in lakes, and their absence indicates a marine origin. Glacial erosion features are present on the basal contact, suggesting that the ice was grounded at times. The Mineral Fork Tillite contains some stratified sandstone and conglomerate, but most of the tillite is poorly sorted, a lithified mixture of matrix-supported rock fragments, cobbles, sand, and mud mixed with sand and silt.

The Mineral Fork Tillite is a glacial deposit that accumulated in a tidal environment at or near sea level. The magnetic signature in the rock records the Earth's magnetic field at the time of accumulation and indicates that the tidal environment was near the equator. There was a tidewater glacier at sea level near the equator, a circumstance that is not present on the Earth today. Deposits like the Mineral Fork Tillite and of similar age occur on every continent regardless of latitude, and the sequence of rocks is similar in each occurrence: glacial deposits, then a laminated carbonate (not present in the Wasatch Mountains, but easily seen on Antelope Island in Great Salt Lake). The story developed by Paul Hoffman of Harvard University and Daniel Schrag (2002) is that almost the entire Earth was ice covered, a snowball Earth. Only a few enclaves of open ocean might have existed. Due to the ice cover on the rocks and the oceans, the chemical weathering system present on the Earth today did not exist. This chemical weathering system together with atmosphere-ocean dynamics mediates the CO_2 (carbon dioxide) content of the atmosphere. In the ice-covered earth, CO_2 from volcanic emissions increased in the atmosphere until CO_2 was high enough in concentration to produce greenhouse warming and melting of the glaciers. Chemical weathering then produced the calcium and bicarbonate in the oceans that precipitated as the laminated carbonate rocks that overlie the tillite.

Stability of the Earth's climate is due to a balance between CO_2 derived from degassing the interior of the Earth through volcanic emissions and consumption of CO_2 in weathering. Too much cooling leads to a runaway ice-albedo feedback: that is, an ice-covered earth reflects (albedo) more of the sun's energy back into space. Hoffman and Schrag show that during the 200 million years preceding the appearance of large multicellular animals 750 to 550 million years ago, fragmentation of Rodinia, a large, long-lived supercontinent, was accompanied by intermittent, widespread glaciation. The breakup of Rodinia exposed new rocks to the effects of weathering. Chemical weathering reactions transferred carbon dioxide from the atmosphere to the oceans. A lowered CO_2 content of the atmosphere resulted in drastic cooling. The ice line even moved to sea level near the equator. Biological productivity in the oceans collapsed for millions of years, and the hydrologic cycle that mediates the carbon dioxide content of the atmosphere was interrupted. Then volcanic eruptions caused the CO_2 content of the atmosphere to increase drastically. When the CO_2 content was 350 times the modern level, the greenhouse effect terminated the glaciation.

The Mutual Formation (Purple Quartzite)

A blanket of red to purple quartzite and red and green shale covers the Mineral Fork Tillite. The deposition of this blanket was controlled by a rise in sea level that occurred at the end of Precambrian glaciation as water was returned to the sea from the ice caps. A snowball Earth episode isolated the oceans from atmospheric oxygen. One consequence was increased reducing conditions and an increase in iron solubility in the oceans. At the close of a snowball Earth episode when the ice cover melted, the oceans were exposed to atmospheric oxygen and the dissolved iron was oxidized reducing its solubility. Precipitation of the oxidized iron may account for the pronounced purple color and iron banding of the Mutual Formation in a marked color contrast with the underlying Big Cottonwood Formation.

Paleozoic Rocks

Cambrian Rocks (545 to 505 million years ago)

A great interval of time is missing between deposition of the Precambrian Mutual Formation and deposition of the overlying Paleozoic sequence of sediment, possibly 500 million years. Entire mountain ranges were removed by erosion during this interval of time. The Paleozoic Era begins with the Precambrian exposed in the center of the North American continent with a great sandy beach to the west. Farther west toward deeper water, mud and limestone were deposited on a passive continental margin. But here in the Wasatch Mountains, the sandstone is overlain by shale, then a thick sequence of limestone. The deposition of these rocks began on a newly formed continental margin as the core of the continent rifted from its parent.

The sequence of rifting events begins with a great continental rift as the continent that was to become North America (Laurentia) broke away from its parent, a portion of Rodinia. A major elongate rift depression formed that was bounded by great normal faults. The surface layers of the Earth were arched upward and pulled apart in response to heating from the mantle. Volcanic rocks, most commonly basalt, extruded as the rifting continued. The continental rift was eventually filled with the ocean as the two continental fragments moved farther apart and new oceanic crust formed between them. As the continental fragments continued to move apart, the edges cooled, became denser, and subsided. The gradual subsidence permitted the accumulation of a thick sequence of sediment, mostly in shallow water on the newly formed continental margin. The accumulated sediment added weight to the continental margin, which contributed to the continued subsidence.

The Tintic Quartzite is the first and oldest sedimentary rock in the Wasatch Mountains in the Paleozoic Era. This quartzite formed from a westward thickening sand wedge that accumulated on a tropical beach. The Tintic Quartzite is a Middle to Lower Cambrian white, gray, reddish brown, faintly flesh, or salmon-colored quartzite up to 2,000 feet thick with a prominent conglomerate at the bottom. A basalt flow is widespread in the formation. The quartzite is texturally mature, and cross beds indicate the directions of current movement. A local basal conglomerate records the action of vigorous current or wave motion. The Tintic Quartzite is stratigraphically the youngest eastern featheredge of a westward thickening, shallow marine quartz sand wedge. In other words, this unit thins eastward, and becomes a near-shore beach sand. Cambrian strata in the Wasatch are one-tenth the thickness of strata in western Utah. The great thickness in the west accumulated as the slow loss of heat and accumulated sediment thickness caused subsidence of the rifted margin of North America.

The Tintic Quartzite is resistant to erosion and well exposed wherever it occurs. The Tintic Quartzite is approximately equivalent to the Tapeats Sandstone and Bright Angel Shale exposed in the Grand Canyon of Arizona or the Pioche Formation in western Utah and Nevada.

The Ophir Shale consists of three separate beds. The lower micaceous shale is olive to bluish green, 240 feet thick, and directly overlies the underlying quartzite. The middle member is limestone about 80 feet thick with yellowish silty lamina. The occurrence of rich ore bodies of silver and lead in this Middle Ophir Limestone in many of the mining districts of Utah was well known to the early prospectors and miners. The Ophir Limestone is thoroughly prospected with pits and small mine openings. The upper Ophir is sandstone that is cemented with calcite. The Ophir Shale is equivalent to the Muav Limestone in the Grand Canyon. At the time of deposition of this Cambrian shale and limestone, Utah was near but slightly north of the equator.

The Maxfield Limestone of Middle Cambrian age consists of an upper dark gray oolitic dolomite, middle mottled dolomite, limestone, nodular shale, and a lower massive gray mottled dolomite and limestone with yellowish silty lamina. A creamy white dolomite about 15 feet thick is a prominent and useful marker at the top of the Maxfield. The Maxfield occurs only in the Wasatch with maximum thickness of 1,200 feet north of City

Creek Canyon. The thickness varies greatly due to the erosional and angular unconformity at the top. The Maxfield is equivalent to the Lodore Formation in Dinosaur Monument near Vernal, Utah, east of the Wasatch Mountains. Cambrian limestone accumulated in a broad, subtidal platform shoaling to the east with deeper water slope deposits to the west in western Utah and Nevada.

No Ordovician sediments (505 to 438 million years ago) accumulated east of the Mesozoic thrust belt and thus none are present in the Wasatch Mountains. Silurian (438 to 408 million years ago) and Devonian (408 to 360 million years ago) sediments are also missing from the Central Wasatch Mountains. Some sediment of these ages may have been deposited, but they were later worn away by erosion for there is an erosion surface on the Cambrian. The Cambrian-age Maxfield Limestone has been removed by erosion in places, so Mississippian rocks rest directly on the Cambrian Ophir Shale.

Mississippian Rocks (360 to 320 million years ago)

In Mississippian time (360 to 320 million years ago), the shoreline of the continent was just north of the equator and shallow, tropical seas covered the continental shelf. The Mississippian ocean that represents the greatest limestone-producing interval of geologic time covered the entire state of Utah. The western margin of the North American continent had ceased to be a passive margin. It was involved in active mountain building resulting in the construction of the Antler mountain belt shown in Figure 9. The sedimentary rocks that accumulated then are mostly carbonates. The Mississippian carbonate rocks formed in the Antler fore-land basin that extended from Nevada through Utah to the transcontinental arch to the east. They have been divided into a series of separate formations based on subtle differences in the rocks. The tropical climate near the equator favored the growth of reef-building organisms. Water depth and depth fluctuations were critically important in determining the nature of the sediment deposited. Limestones are generally formed in shallow water. Even though the limestones were deposited in a tropical to subtropical climate, glaciation in polar regions had an effect, because glacial ice stored great quantities of water and caused sea level to fall and sand and silt to encroach upon the limestones.

The first in the Mississippian series is the Fitchville Formation, followed by the Gardison Limestone, then the Deseret Limestone and the Humbug Formation. The Deseret and Gardison cannot always be distinguished and are often mapped together as a single unit. The top of the Mississippian and perhaps even extending into the Pennsylvanian is the Doughnut Formation.

The Fitchville Formation is mostly dolomite (calcium-magnesium carbonate) in the Wasatch Mountains. The Fitchville Formation unconformably overlies the Cambrian in places in direct contact with the Ophir Shale where the Maxfield has been removed by erosion, but in other places it is in contact with the erosion surface on the Maxfield. A bed of sandstone marks the base of the Fitchville. A layer of white-weathering dolomite about three feet thick occurs at the top. The Fitchville is equivalent to the Pilot Shale of western Utah that is composed of detritus from the Antler Mountain belt in Nevada. The Fitchville Formation is host to most of the rich, lead-silver replacement ore bodies in the Little Cottonwood Mining District.

The Gardison Limestone, named for a spur in the northern East Tintic Mountains, is a shallow-water, medium to dark gray, cherty and noncherty limestone. The basal bed is cross-bedded and fills channels cut into the Fitchville Formation. The Gardison Limestone formed from rapidly accumulating fragments of shells and skeletons of living organisms on a carbonate shelf. The top 16 feet is dark gray, deeper-water limestone.

The Deseret Limestone, described in its type locality on the west side of the Oquirrh Mountains by James Gilluly (1932), consists of three members. The bottom member is a dark, organic phosphatic shale, siltstone and mudstone enclosing large calcareous concretions. Pelletal and oolitic phosphorite is also present. The middle member formed in moderately deep to shallow water and consists of slope-deposited siltstone, sandstone, and cherty limestone. The upper member is a prograding carbonate buildup. The Deseret is equivalent to the Redwall Limestone that forms prominent red-stained cliffs in the Grand Canyon.

Alternating beds of sandstone and limestone, reflecting rising and falling sea level, characterize the Humbug Formation. The sea level fluctuations are related to glacial episodes on Pangea in Mississippian time.

The Doughnut Formation consisting of black shale and gray to black limestone is about 430 feet thick. The Doughnut Formation is rich in organic matter and may be a petroleum source bed.

Pennsylvanian Rocks (320 to 286 million years)

The Pennsylvanian period lasted 34 million years, from 320 to 286 million years ago. A great thickness of sediment, more than 3 miles thick, accumulated west of the Wasatch Mountains in the Oquirrh basin, but the Weber Quartzite and Round Valley Limestone together represent not more than 2,100 feet, accumulated in the same time period in the Wasatch. The Round Valley Limestone (Lower Pennsylvanian) and the Weber Quartzite (Upper Pennsylvanian) record a sequence of falling sea level from open marine Round Valley Limestone upward to near-shore sand and limestone of the Weber Quartzite in the Wasatch Mountains. The Weber is a wind-deposited sand dune deposit in eastern Utah. The Weber Quartzite is equivalent to the Esplanade Sandstone in the Grand Canyon.

The type locality of the Weber Quartzite in the northern Wasatch Mountains is a sandstone or quartzite with 15 to 20 percent limestone and dolomite beds 1,970 feet thick. The sandstone and quartzite have small- to medium-scale cross beds. The Weber Quartzite is considered to have accumulated in a marine environment and is transitional to the Oquirrh Formation to the west. Harold Bissell and Orlo Childs (1958) describe the Weber in eastern Utah. The Weber Quartzite is friable, fine-grained sandstone that is 1,200 to 1,500 feet thick. Large-scale cross beds indicate an eolian origin. Some intercalated fluvial sandstone, siltstone, and shale are present. The eolian Weber intertongues with the fluvial Maroon Formation (Maroon Bells high peaks near Aspen, Colorado). Dunes prograde toward the Uncompahgre in Colorado.

The Weber Quartzite is equivalent to the enormously thick Oquirrh group to the west and the Paradox Formation in southeastern Utah. Fluctuations in sea level produced cyclic sedimentation in the Oquirrh and Paradox Basins. These fluctuations are interpreted to be the result of glaciation near the polar portions of Pangea.

The Weber Quartzite and age-equivalent Oquirrh and Paradox Basin sediments accumulated as Pangea came together with South America and Africa in Pennsylvanian time. The plate collisions formed a series of basins in Utah and uplifts in Colorado known as the Ancestral Rockies.

Permian Rocks (286 to 248 million years ago)

The top of the Permian is an unconformity (missing interval of time) of about 5 million years' duration that is coincident with the greatest mass extinction of life on earth of all time. At the top of the Permian 248 million years ago, life in the oceans nearly disappeared; 85 percent of sea species and 70 percent of vertebrate genera living on land vanished in less than 5 million years. Meteorite fragments have been found in Antarctica that correspond in time to the extinction event. The presence of meteorite fragments of this age suggests to some investigators that a meteorite impact played a role in the extinctions.

In the Wasatch Mountains, the Park City Formation consists of limestone, limy shale, and sandstone. Part of the Park City Formation, the Phosphoria, accumulated in an unusual deeper water environment of upwelling ocean currents. In southern Utah the Permian is arid, shallow, seasonal lakes and sand dunes. The Park City Formation in the Wasatch was a common source of limestone for the manufacture of lime in lime kilns that are located in Mill Creek Canyon about 0.7 miles from the mouth of the canyon and at the mouth of Rattlesnake Gulch (Keller 2001). The Park City Formation is the marine equivalent of the Kaibab Limestone on the rim of the Grand Canyon.

Mesozoic Rocks

Triassic Rocks (248 to 206 million years ago)

Triassic time records a major change in sedimentation from marine to continental. The lower Triassic sediments are marine, the middle is missing, and the upper Triassic sediments are continental.

The Lower Triassic Woodside Shale is 700 to 800 feet of marine, dark red shale, siltstone, and fine sandstone with thin white limestone beds. The Lower Woodside Shale contains gypsum (calcium sulfate). The Woodside Shale is never very well exposed. In most of the occurrences in the Wasatch Mountains, it is covered with talus or vegetation, and few good outcrops are present. This unit is equivalent to the bottom of the Moenkopi Formation in southeastern Utah.

The Thaynes Formation is shale, fine-grained sandstone, and limestone. The thick limestones reveal the folding on the mountain front of Grandeur Peak and on the north side of Mill Creek Canyon. The Thaynes Formation accumulated in slightly deeper water than the Woodside and also correlates with the Moenkopi of southern Utah.

The Ankareh Formation is brilliant red shale with a gritty quartzite in the middle. The grit, known as the Gartra Grit, is an unusually light purple to red color sandwiched between two red shales. The Ankareh Formation consists of continental red beds that conformably overlie the marine Thaynes Formation and are in turn conformably overlain by the Nugget Sandstone. An excellent exposure of the Gartra Grit and the adjacent shale beds is in the road cut where Interstate-80 and Interstate-215 meet. The Ankareh Formation was deposited near the eastern shore of the Cordilleran ocean on a broad, gently sloping plain. The sediments were derived from the Uncompahgre Uplift (part of the Ancestral Rockies formed in Pennsylvanian time) transported in westward- and northward-flowing streams. The Ankareh was deposited by the Chinle-Dockum paleo river system. Detrital zircons show that the sediment originated in the core of the Ancestral Rocky Mountains and the Amarillo-Wichita uplift in Texas. Zircons are found in these Triassic beds with consistent ages of 515 to 525 million years old all the way from Texas to Nevada. This river system preserves the remains of many forests on the edge of the Pangea continent 225 million years ago.

The Petrified Forest National Park in Arizona is in rocks equivalent to the upper Ankareh. These rocks contain great numbers of petrified conifer logs. The silicification that preserved the logs may be related to bentonite (volcanic ash) beds in the formation.

The lowermost Ankareh was deposited on tidal flats. Meandering stream deposits that make up the bulk of the formation covered the tidal flats. Some estuarine facies were present at the boundary between meandering stream and tidal flat. The sediments deposited by the meandering streams have characteristics that suggest that the streams were overloaded with sediment. Tectonic rejuvenation of the Uncompahgre uplift, first formed in Pennsylvanian time, could have supplied the excess sediment. Eolian facies are present at the top near the Nugget contact. The lower member of the Ankareh is equivalent to the upper Moenkopi of the Colorado Plateau. The Gartra Grit Member is equivalent to the Shinarump Member of the Chinle, and the upper portion of the Ankareh is equivalent to the remainder of the Chinle. Brandley (1990) described the Ankareh Formation in detail. Utah was located at latitude $10°N$ at the time of deposition of the Ankareh and the shallow, continental shelf basin extended from Utah to the central Nevada Sonoman mountain belt.

A major faunal mass extinction occurred at the Triassic-Jurassic boundary 206 million years ago. The mass extinction was the third largest in earth history: 30 percent of marine genera and 50 percent of tetrapod species were lost, and there was a turnover of more than 95 percent of megafloral species. The climate also warmed by as much as $3°$ to $4°C$ probably because of an increase in greenhouse gases such as CO_2 in the atmosphere.

Jurassic Rocks (206 to 144 million years ago)

The Nugget Sandstone together with its equivalent Navajo Sandstone in southern Utah and Aztec Sandstone in southern Nevada are part of the greatest sea of windblown sand that ever existed on earth. The sea of desert sand covered Utah, parts of Wyoming, Nevada, Colorado, Arizona, and New Mexico. The Nugget Sandstone is composed of fine to medium-size grains of sand and the color varies from pale pink to pale orange. In southern Utah, the equivalent Navajo Sandstone is often deep red to brown, but may be bleached to white due to the flow of subsurface fluids that contained hydrocarbon. The sand accumulated in a foreland bulge that gradually subsided in response to lithospheric plate convergence on the west and formation of the Nevadan Mountains. The rate of subsidence eventually exceeded the supply of sand so that the vast desert sand sea was submerged and covered with marine sediments.

The Twin Creek Limestone consists of several units. The bottom Gypsum Spring Member includes red beds, brecciated limestone, and chert. This initial deposit of the Twin Creek formed in a very shallow sea advancing eastward. Extensive masses of gypsum over a large area indicate a hot and arid landscape. Then conditions changed for deposition of the next layer. The Slide Rock member formed in shallow ocean water of normal salinity. Succeeding overlying layers also formed in a shallow marine environment deepening westward. The Twin Creek Limestone is the youngest sedimentary rock unit in the Tri-canyon area of the Central Wasatch Mountains. To see younger units you must go east to see Cretaceous-age rocks, south and east to see the younger Jurassic Morrison Formation, and northeast to see the Preuss Formation. These formations record the last vestige of an ocean in the Wasatch Mountains.

The Twin Creek marine basin intervened between the emergent North American continent and the Nevadan phase of Cordilleran Mountains to the west. The mountain building involved extensive thrust faulting. The thickened stack of thrust sheets to the west loaded the crust, causing a downward bulge and foredeep basin in which the Twin Creek and later rocks accumulated.

Cretaceous Rocks (144 to 65 million years ago)

Coarse conglomerates are among Cretaceous rocks in the Wasatch Mountains. Vigorous mountain streams that once flowed eastward from high mountains to the west deposited these conglomerates. The coarse conglomerates are named the Kelvin, Frontier, Henefer, and Echo Canyon Conglomerates. They are exposed east of the Wasatch front in Mountain Dell Canyon and Echo Canyon, for example. The conglomerates were deposited at the foot of mountains and on the margin of the Cretaceous foredeep basin. Coarse conglomerates grade eastward into flood plain, shoreline, and marine deposits. The shoreline and flood plain deposits contain the rich coal deposits of Carbon County.

Cenozoic Rocks

Tertiary Rocks (65 to 1.8 million years ago)

Various mountain-building events that have affected this region produced a series of intrusive igneous rocks. A sweeping arc of magmatism formed a belt of igneous rocks known as the Wasatch Igneous Belt. Thomas Vogel and coauthors (1998) described the western group of intrusive rocks in the Wasatch Igneous Belt that includes the Little Cottonwood, Alta, and Clayton Peak intrusives in Big Cottonwood and Little Cottonwood Canyons. All of these intrusives are high in potassium, calcium, and silicon, and depleted in iron, and are typical of igneous rocks formed on a convergent continental margin. The Little Cottonwood igneous intrusive is the youngest and the most silicic (granite to granodiorite) and was intruded at 30.5 ± 0.5 million years ago. The Alta stock (intruded 34.3 ± 1.5 million years ago) is granodiorite to quartz diorite (contains less silica) with biotite and minor hornblende (iron and magnesium bearing dark-colored minerals). The Clayton Peak stock (intruded 35.5 ± 1.5 million years ago) is the most mafic (the most iron and magnesium and the least silicon). The Clayton Peak stock is quartz monzodiorite to quartz monzonite containing the minerals clinopyroxene, orthopyroxene, and biotite (dark-colored minerals containing iron and magnesium).

The three stocks cannot be related by differentiation of a parent homogenous magma, as they did not all come from the same magma chamber. The origin is likely partial melting of a mafic to intermediate crust. The crust was first thickened during mountain-building events, then the thickened crust began to extend and thin. Hot mantle upwelled into the shallower mantle and lower crust and began melting the surrounding rocks. Iron- and magnesium-rich magma formed in the mantle intruded into the lower crust, and this hot magma began melting the Precambrian crystalline crust. These igneous rocks come from melted crust and not from the mantle.

Mountain-Building Events

The landform that we call the Wasatch Mountains is a complex of many mountains of different ages extending back in time to the beginning of the North American continent and beyond to the fragments of continent that were to be assembled to form North America. The present-day Wasatch Mountains are youthful and uplift continues today, but the bedrock exposures record mountains that were eroded away eons ago. The oldest mountain-building events preserved in the rocks of the Wasatch Mountains predate the formation of the Wasatch and in fact predate the assembly of the North American continent.

Mountain building is one response to the ever-changing distribution of continents and ocean basins on earth. Laurentia, the Precambrian core of North America, was a part of two super-continents and has been involved in a number of mountain-building events.

Precambrian crystalline basement in Utah is the root of ancient mountains. The metamorphism and the igneous intrusion in crystalline basement are produced by the pressure and weight of the mountains. Mountains were being made long before the North American continent took its present shape. The oldest episode of mountain building in the Wasatch Mountains is recorded in the roots of mountains that are preserved in the Precambrian metamorphic rocks. Additional mountains must have been produced when the crystalline basement of central Utah was accreted to the Wyoming Craton of northern Utah. Those mountains have long since been eroded away, but their roots are exposed in the Wasatch Mountains.

The Farmington Complex, near the inferred margin of the Archean craton, preserves the oldest evidence of mountain building in the Wasatch. Layered metamorphic rock of Archean age (more than 2,500 million years old) consists of interlayered biotite-hornblende quartz monzonite gneisses and schists. Metamorphosed plutonic rocks of Early Proterozoic age north of Weber Canyon are biotite-hornblende quartz monzonite gneiss. Amphibolite formed from basalt is a prominent constituent. Younger Precambrian mountain-building events in Utah were associated with accretion of each of the blocks of Precambrian terrain shown in Figure 6 at 1,660 to 1,700 million years ago and 1,760 to 1,800 million years ago.

The margin of the continent developed in latest Precambrian to latest Devonian following rifting from Rodinia. A nearly continuous succession of Proterozoic to Upper Devonian sedimentary rocks accumulated in the western United States, although the western limit of the carbonate shelf disappears beneath later thrust sheets.

From Cambrian to Ordovician through Silurian and most of Devonian time, the western continental margin was a passive continental margin. It was not a lithospheric plate boundary. The plate boundary was off in the ocean somewhere. No mountains were formed on the developing western continental margin throughout the early part of Paleozoic time. But then things changed. The western continental margin became a convergent lithospheric plate boundary in Mid-Devonian to Mississippian time when the first mountain-building event in the Paleozoic resulted in the accretion of the Antler island arc. This is the first in a sequence of eight mountain-building events that are summarized in Table 2. This event took place far to the west of the Wasatch in Nevada, and the faulting and other mountain-building effects are not exposed in Utah.

In Pennsylvanian time, another major mountain-building episode, the second in the series, resulted from events far from Utah on the southeastern margin of North America. The big supercontinent of Pangea was being assembled. The big continent that we now know as South America was accreted to North

Event Number	Mountain-Building Event	Age	Location	Expression in the Wasatch
1	Antler	Late Devonian to Early Mississippian	Central Nevada	Erosion or nondeposition of Devonian sediment
2	Ancestral Rockies	Early Pennsylvanian	Central Colorado to Eastern Utah	Weber Quartzite on margin of the Oquirrh Basin
3	Sonoman	Late Paleozoic to Early Mesozoic	Central Nevada	Park City Formation, Woodside Shale
4	Nevadan	Early to Mid-Jurassic 163 to 152 million years ago	Western Nevada	Nugget Sandstone
5	Eureka–Central Nevada	Late Jurassic	Central Nevada	igneous intrusive rocks at Gold Hill, Pilot Range, and Newfoundland Mountains
6	Sevier	Cretaceous 145 to 75 million years ago	Central Utah	Folding and thrust faulting
7	Laramide	Early Cenozoic	Eastern Utah and Colorado	Uinta arch westward to Big Cottonwood uplift
8	Basin and Range	15 million years ago to present	Eastern Nevada and Western Utah	Wasatch Fault

Table 2. *Summary of mountain-building events*

America in a major collisional event that produced a major mountain chain on the southeast coast of North America and the Ancestral Rocky Mountains in Colorado shown in Figure 10. The Uncompahgre uplift in southeastern Utah is part of the Ancestral Rockies.

This event also resulted in the formation of two marine basins: the Oquirrh basin that covered much of western Utah and the Paradox Basin that covered southeastern Utah. Both of these

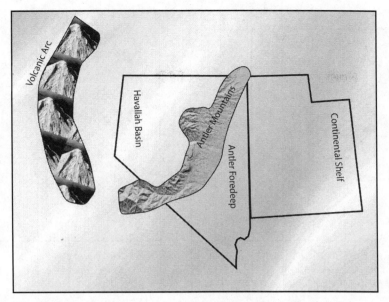

Figure 9. *The Antler mountain-building event, the Antler foredeep basin, and continental shelf in Utah. The Antler mountain-building event took place in Late Devonian to Early Mississippian time.*

basins are fault block basins. These basins subsided and filled with sediment more or less keeping pace with the subsidence. More than 20,000 feet of sediment accumulated in the Oquirrh Basin. The Paradox Basin was less deep and accumulated less sediment. The Paradox Basin was more restricted to the circulation of the open ocean, and evaporating seawater deposited a thick layer of sodium chloride and other associated salts in the Paradox Salt Formation.

The third mountain-building event began to form in Permian time and continued on into the Triassic. The Oquirrh and Paradox Basins had filled. Sand was blowing around on top of the Paradox Basin sediments now exposed in the monuments we see in Monument Valley. The western margin of North America was still a convergent lithospheric plate margin and it accreted the terrain in Nevada that we know as Sonomia, shown in Figure 11. The eastern edge of accreted Sonomia is a thrust fault wedge

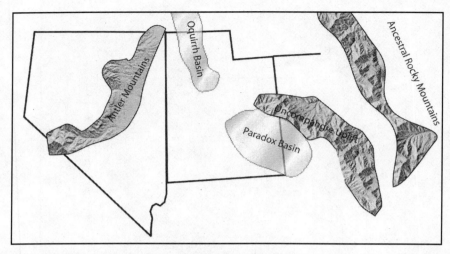

Figure 10. *The location of Pennsylvanian Ancestral Rockies and the Oquirrh and Paradox basins in Utah. The Ancestral Rockies formed in Early Pennsylvanian, and the basins lasted to Early Permian.*

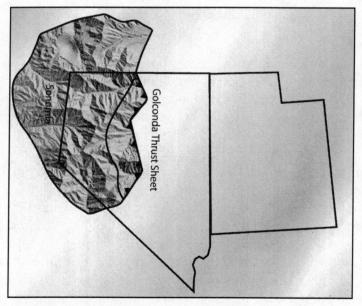

Figure 11. *The Sonoman mountain-building event and the accreted terrain of Sonomia. The Sonoman is a Late Paleozoic to Early Mesozoic event.*

representing the portion of Sonomia that was pushed up onto the continental shelf of North America, thus thickening the crust there. The Sonomian mountain-building event together with accreted terrain and faulting are all located in Nevada. The Park City Formation and the Triassic Woodside Shale and Thaynes Formation represent this mountain-building event in the Wasatch Mountains.

Three Mesozoic to Tertiary-age events are next in the sequence of mountain building. Prior to Mesozoic and Cenozoic deformation, Archean basement was overlain by a thick sequence of continental shelf sediments of Upper Proterozoic to Lower Cambrian clastics and Paleozoic carbonates. The mountain building affected great thicknesses of Early Paleozoic limestone, shale sandstone west of the Wasatch Mountains. The mountain building thrust the sediments eastward.

The series of three episodes of mountain building, numbers four, five, and six, began in the Early Jurassic. These episodes of mountain building are called the Cordilleran series. The three stages of Cordilleran mountain building are shown in Figure 12. The effects moved eastward through Nevada in Jurassic time and into Utah in Cretaceous time. Utah was most affected in the Cretaceous and was bypassed for Colorado in Late Cretaceous to Tertiary time.

The three Cordilleran events took place from Jurassic to Tertiary time and produced a series of thrust-faulted mountain belts in Nevada and Utah. A typical mountain belt consists of, from west to east, a thrust belt, a foredeep basin, a forebulge high, and a back-bulge basin. The thrust belt consists of a thick wedge of stacked thrust plates. A single thrust plate may be up to 50,000 feet thick, and thick stacks would have formed mountains similar to the Andes. The enormous weight of the stacked thrust sheets caused the crust to sag and form the foredeep basin that could accumulate more than 10,000 feet of sediment.

As the mountain-building events migrated eastward, the thrust belt, foredeep basin, forebulge high, and back-bulge basin also migrated eastward. By Late Jurassic, the back-bulge basin had migrated east of Utah into Colorado, and Utah was mostly a

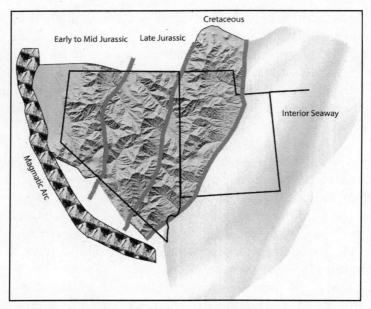

Figure 12. *Three stages in the development of the Cordilleran mountain building. First the Early to Mid-Jurassic Nevadan event in Nevada (163 to 152 million years ago), then the Central Nevada–Eureka thrust belt formed in Late Jurassic, and finally the Cretaceous Sevier event in Utah took place in Late Cretaceous time (145 to 75 million years ago).*

forebulge high. Erosion of the high resulted in an unconformity between the Morrison and the Cedar Mountain and Kelvin Formations. These formations are not present in the Central Wasatch Mountains.

The first mountain building of the Cordilleran, the fourth in the series, is a Jurassic (208 to 144 million years ago) event that produced many of the features of the Sierra Nevada Mountains— not the present topography, but many of the rock characteristics— so it is called the Nevadan mountain-building event. The Nevadan mountain building formed the structure of the Sierra Nevada Mountains, but the uplift apparent today occurred much later in Pliocene-Pleistocene time (5.1 million to 10,000 years ago).

The western interior seaway that ultimately connected the Gulf of Mexico with the Gulf of Alaska in the Cretaceous first formed in Jurassic time. The compression and weight of the thickened crust of the Jurassic mountains west of Utah formed a foreland marine basin. Deposits in the Middle Jurassic back-bulge basin define a broad, shallow basin in Utah first covered by a shallow ocean, tidal flats, flat evaporating pans, or coastal sand dunes represented by the Twin Creek Limestone and Preuss Formation in northern Utah, the Arapien and Twist Gulch in central Utah, and the Entrada, Curtis, and Summerville in southeastern Utah. The Jurassic back-bulge basin was later covered by near sea-level flood plains of the Stump and Morrison Formations, neither of which are present in the Wasatch Mountains.

Streams were flowing eastward across Utah into the Sundance Sea from the Nevadan Mountains out to the west. Utah at times was a part of the shoreline of the Jurassic Sundance Sea and at times it was under water. Some of the sediments in the Wasatch Mountains in Jurassic time are not marine sediments; instead they are sediments that accumulated on land. For example, the extensive Jurassic Nugget Sandstone is a windblown sandstone that accumulated next to the Sundance Sea. The Morrison Formation, exposed in Dinosaur National Monument near Vernal, consists of coarse gravel and sand that came from the Nevadan Mountains in streams that flowed eastward into the Sundance Sea. The Morrison Formation also contains volcanic ash, evidence for volcanic activity of the time.

Mountain building progressed across Nevada to the Eureka–Central Nevada mountain belt, the fifth event. The thrust faults and folds of the Late Jurassic affected rocks that were deposited on the ancient Precambrian continental crust of North America. Some evidence of this deformation is present in Utah. Igneous intrusions in the Newfoundland Mountains, Pilot Range, and Gold Hill in the Great Salt Lake Desert of western Utah date from the Late Jurassic.

The sixth mountain building, the Sevier, took place in Cretaceous time and finally reaches Utah with major effects. Thrust faulting in Utah began in earnest in latest Jurassic to earliest Cretaceous and reached its zenith during Late Cretaceous. The thrust fault sheets are composed mostly of the sedimentary rocks that had previously accumulated on the Precambrian basement rocks. Many plates of sedimentary layered rocks in the sequence of up to five stacked thrust sheets were pushed eastward 25 to 30 miles and sometimes up to 50 miles. The plates were extensively folded and faulted, as can be seen in Parleys Canyon folds and the overturned strata at Mount Nebo near Nephi, Utah.

The thrust faulting continued into middle to late Eocene as the thrusting declined gradually and ended by 50 million years ago. The thrusting overlapped the Laramide mountain-building event, the seventh in the series.

The main distinction between the Sevier and Laramide mountain-building events is the thickness of crust affected. The Sevier event affected mostly the thinner sedimentary cover while the Laramide event penetrated deeper into the Precambrian crystalline basement rocks.

The Cretaceous-age Sevier mountain building was associated with the accretion of some terrain to the continent. The exact terrain accreted is as yet uncertain. The lithospheric plate convergence was oblique, that is, North America was not converging at right angles with the Pacific Ocean plate. There was a component of force that produced right lateral strike slip motion of the accreted terrain. The accreted material moved to the north, and may be a part of British Columbia, Alaska, or the Blue Mountains in Oregon.

The Sevier event is associated with a series of thrust faults in Utah. They began with the Willard, and the Charleston Nebo thrust fault and the associated folding that is so prominently displayed in our Wasatch Mountains. Thus, the thrust faults in the Wasatch Mountains are Cretaceous. The thrust faults in the area covered by this guide are known as the Alta-Grizzly, Mount Raymond, and Charleston-Nebo thrust faults. The folds in the Wasatch Mountains, the Parleys Canyon syncline, the Spring Canyon anticline, and the Emigration Canyon syncline are also Cretaceous in age. There is a small intrusive igneous rock in the Wasatch Mountains at Argenta, near Butler Fork in Big

Cottonwood Canyon, that is Cretaceous in age. There are two or three other small intrusives there, but there is not a lot of igneous activity in the Wasatch Mountains associated with Cretaceous mountain building. Most of the Cretaceous igneous activity is far to the west in the Sierra Nevada Mountains.

The Cretaceous mountain-building event did not generally involve deep levels of the Precambrian crystalline basement. Paleozoic and Mesozoic sedimentary rocks were folded and faulted, but Precambrian rocks were not affected very much. The Willard thrust sheet exposed near Willard Peak is in the Paleozoic sediments. Precambrian basement rocks are prominently exposed in the Wasatch Mountains there because of the displacement on the Wasatch fault, but the Willard thrust is higher up in Paleozoic rocks. The Charleston Nebo thrust down on Mount Nebo, expressed there as a large, recumbent anticlinal fold, also affects Paleozoic and Mesozoic rocks. Therefore the Sevier is referred to as a thin-skinned event. The mountain range produced after the Jurassic Mountain range in Nevada preserved the western interior seaway. Deposits of volcanic ash accumulated in the seaway from the volcanic activity that accompanied this mountain-building event.

The Cretaceous Sevier event is followed by the seventh event. The seventh event, the Laramide, named for the Laramie Mountains in Wyoming, is a Cretaceous to Tertiary event that followed right on the heels of the Sevier event. Laramide-age uplifts in Utah and Colorado are shown in Figure 13. The Laramide event was involved in formation of the Wind River Mountains and the Big Horn Mountains in Wyoming, and the Uinta Mountains in Utah. The core of the Wind River Mountains consists of old Precambrian rocks that are part of the core of the ancient Wyoming continent. The Laramide event affected Precambrian rocks in each of these mountain ranges and so it is called a thick-skinned event. In Utah, additional Laramide structures include the San Rafael Swell and the Waterpocket Fold at Capitol Reef, both of which are cored by faults in the Precambrian basement.

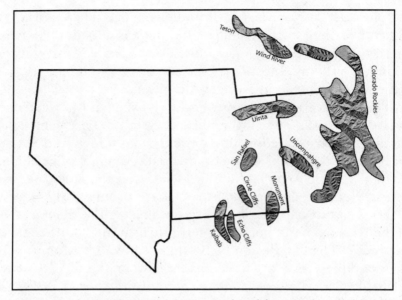

Figure 13. *The distribution of the Laramide uplifts in Utah and Colorado. Laramide structures formed in the Early Cenozoic.*

The thick-skinned deformation that carried eastward into the Rocky Mountains of Colorado was produced by a low inclination or flat-dipping, subducting slab of Pacific Ocean lithosphere. The flat dip was produced by an increase in the rate of convergence and by subduction of younger, hotter, more buoyant oceanic lithosphere. It wasn't until the subducting slab reached Colorado that it was deep enough to generate igneous rocks. The flat-dipping slab also uplifted the base of the interior seaway, causing the sea to recede. All the streams continued to flow eastward from the Cretaceous Mountains located west of Utah. The streams then flowed into large lakes that replaced the interior seaway. Remnants of the lake deposits are preserved at Bryce Canyon and Cedar Breaks. The effects of Laramide deformation are seen in the Wasatch as a westward continuation of the Uinta Mountains into the Wasatch that is known as the Big Cottonwood uplift centered near Little Cottonwood Canyon.

The final episode in the age of the Wasatch Mountains is Basin and Range extension, the eighth event. The geology of the Basin and Range is described in detail by DeCourten (2003). The thickened crust produced by earlier Jurassic, Cretaceous, and Tertiary mountain-building events collapsed and began to extend westward. Kurt N. Constenius (1996) described a rapid drop in North America–Pacific Plate convergence rate and steepening of the oceanic slab. The result was a large reduction in east-west compression and the Sevier mountain wedge collapsed. Total shortening that produced the thickened crust from 150 to 50 million years ago was 65 to 85 miles. Extension due to collapse of the mountain wedge was 155 miles.

Basin-and-Range mountain building differs remarkably from the earlier mountain-building events. The mountain building is the result of extension and normal faulting. The rocks were not extensively folded and no thrust faults were formed during this event, but some very low-angle extensional faults were formed. Normal faults are responsible for the elevation of most of the present-day mountains above the valley floors.

The reasons for the collapse and extension include a thickened crust with a hot, weakened base and release of the compressive forces in the west as the San Andreas Fault formed. The western margin of North America was converted from a convergent lithospheric plate margin to a transcurrent fault margin, the San Andreas Fault. The forces on the San Andreas Fault were insufficient to maintain the mountain elevation. Like a house built on a weak foundation, the house is now collapsing because the forces that kept the mountains elevated have been relieved. The collapse took place on north-south-trending normal faults including the Wasatch fault, which is still active.

Modern global positioning satellite measurements show that the Basin and Range continues to extend and the Basin and Range mountain-building event is still in progress. The distribution of Basin and Range mountains in Utah is shown in Figure 14.

The question of the age of the Wasatch Mountains can now be answered. This latest, Basin and Range mountain-building event produced the topographic elevation that we admire so much.

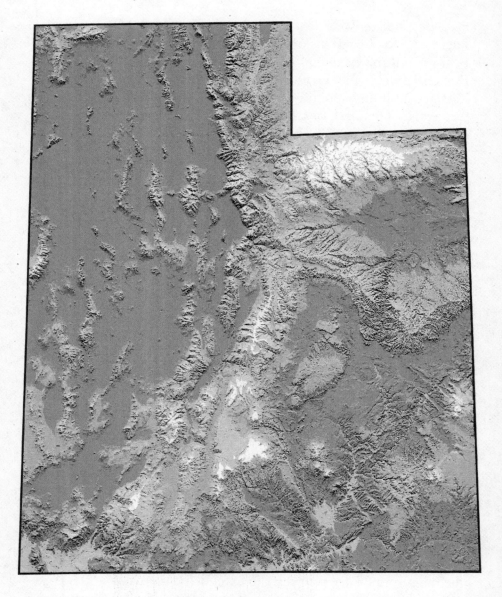

Figure 14. *The present-day configuration of mountain ranges in Utah. These mountain ranges formed during the last Basin and Range mountain-building event that continues today.*

The actual topography is number eight in the series of recorded mountain-building episodes, and is in progress today. Rivers and glaciers eroding the youngest of the mountain-building events produce the detailed erosion of the topography.

Erosion and sculpting of the Wasatch Mountains has exposed some of the structures produced by earlier mountain-building events. The oldest structure is a thrust fault that has a sinuous pattern on the map. The hanging wall side of the thrust fault is exposed in Mill Creek Canyon, and the footwall side of the Mount Raymond thrust fault is exposed on the divide between Mill Creek Canyon and Big Cottonwood Canyon. Hanging wall and footwall sides of a fault are illustrated in Figure 15. The thrust fault exposure reaches the valley at the mouth of Neffs Canyon. The thrust fault has a sinuous pattern because it is a flat-dipping thrust fault and the exposure in outcrop is determined by topography. When erosion proceeds through the hanging wall down into the footwall, the fault is exposed around the ridges and basins. Erosion has exposed the footwall rocks on the flanks of the Cottonwood uplift that is cored by the Precambrian Big Cottonwood Canyon Formation. Layered rocks dip northward on the north flank and curve around and dip northeastward on the flank of the Big Cottonwood uplift.

Folds are also associated with the compressional tectonic event that produced the thrust fault. The folds shown in Figure 2, including the Emigration Canyon syncline, Spring Canyon anticline, and Parleys Canyon syncline, were produced during thrust faulting, together with the prominent folds of the Thaynes Formation on the Grandeur Peak mountain front and in Mill Creek Canyon.

Prominent normal faults exposed in the Wasatch are younger than the folds and thrust faults. The Silver Fork fault is a west-dipping normal fault, and west of the Silver Fork fault is an east-dipping normal fault known as the Superior fault. Looking north from the top of Mount Baldy (Figure 16), the east-dipping Superior fault and thrust sheets can be seen where Tintic Quartzite was thrust directly on top of Mississippian limestone at the Hellgate Cliff just to the north of Snowbird. The great west-dipping

Wasatch normal fault that has elevated the mountain block to expose its geological structure is located at the mount of Little Cottonwood and Big Cottonwood Canyons.

As displacement on the Wasatch rock elevated the footwall, the footwall block was rotated eastward. Approximately six miles of material have been eroded from the footwall at the western edge, but back toward the east erosion is much less, and the Tertiary ground surface is exposed.

4

The Landscape Takes Shape

The Wasatch Fault

The Central Wasatch Mountains lie in the footwall of the Wasatch Fault. Todd Ehlers (2001) infers that the Wasatch fault has been active for 10 to 12 million years, with a maximum exhumation and uplift rate of a few hundredths of an inch per year. The Wasatch Fault is near the easternmost fault of the Great Basin. In the early days of geological investigation of the region, the Great Basin was interpreted as mountain ranges that were crests of broad anticlinal folds. In other words, the basins and ranges were all due to folding. But G. K. Gilbert, in his *Studies of Basin-Range Structure,* published posthumously in 1928, correctly interpreted the formation of basins and ranges as due to block faulting. The mountain ranges are fault block mountains with the basins on the down-thrown side of the faults.

The Lake Bonneville shorelines that surround many of the basins including the Salt Lake Valley provide the evidence that indicates the absolute motion of footwall and hanging wall of the Wasatch Fault. Post-Bonneville movement of the fault south of Salt Lake Valley in Utah Valley is 50 feet, and the Bonneville shoreline remained unchanged in elevation so the footwall block moved upwards. Glacial moraines at the mouth of Little Cottonwood and Bells Canyons are also faulted by the Wasatch fault, as seen in Figure 15.

Effects of Glaciation in the Wasatch

The region encompassed by the Wasatch Mountains has experienced the effects of several episodes of glaciation reflecting global

Figure 15. *Faulted Little Cottonwood and Bells Canyon moraines looking south from Wasatch Boulevard. The Wasatch fault displaced the western end of the moraines downward (hanging wall) by more than 20 meters. The direction of motion on the fault is indicated with arrows, and the hanging wall (HW) and footwall (FW) are also shown.*

Figure 16. *View of Hellgate cliffs, Mount Superior, and the ridge dividing Big Cottonwood Canyon in the north distance from Little Cottonwood Canyon. View is looking north from the summit of Mount Baldy.*

climate and local latitude. Glacial episodes are: Precambrian, 850 million years ago, 1.2 billion years ago, and 2.2 billion years ago; end of Precambrian, 600 million years ago; Early Paleozoic, 500 million years ago; Middle Paleozoic, 350 million years ago; end of Paleozoic, 250 million years ago; and Pleistocene, about 100,000 years ago with the last advance about 22,000 years ago. The youngest glaciation has the most pronounced expression in the Wasatch Mountains.

The latest glacial episode ended barely 15,000 years ago. It is expressed in the Wasatch in a rich variety of land forms carved by the glacial ice: horns, U-shaped valleys, hanging valleys, cirque lakes, and the chaotic, unsorted and unconsolidated glacial debris deposited at the sides and ends of the glaciers. The timing of glaciation in the Wasatch Mountains is thought to be similar to glaciation of mountains in Wyoming and Colorado. The results of [10]Be (Beryllium 10) dating of glaciers by John Gosse and coworkers (1995) in the Wind River Mountains in Wyoming shows that the maximum advance of the last glacial cycle was 21,700 years ago and lasted for about 5,900 years. The glacier retreated 20 miles by 12,100 years ago.

The climate 14,500 years ago at the close of glaciation was different than today. Mean July temperature was 10 to 11 degrees colder than present. January was 26 to 30 degrees colder. By 13,200 years ago, July was 3 to 4 degrees cooler than present. Temperatures were warmer than present by 10,000 years ago.

As a consequence of the coolness of the climate, a large lake more than 1,000 feet deep occupied the valley from 16,000 to 14,500 years ago (Gilbert 1890). Lakes had been present before this time. At 22,000 to 23,000 years ago, a lake occupied a portion of the valley with the lake level at 4,500 feet above sea level. This lake level is named the Stansbury level. The Stansbury lake fell and then began to rise again. By 16,000 to 14,500 years ago, the lake had risen to 5,090 feet above sea level. This lake level is named the Bonneville level. The Bonneville level reached a spill threshold at Red Rock Pass between Preston and Downey, Idaho. Water spilled out into Marsh Creek and flowed into the Snake River. As the threshold was eroded, the lake level fell catastrophically by 350 feet to 4,740 feet above sea level and remained there from 14,500 to 12,500 years ago. The lake level was controlled by the elevation of the threshold at Red Rock Pass. This lake level is named the Provo level. The lake has undergone several fluctuations in level since then because of climate control. The lake fell to 4,250 feet above sea level at 10,000 to 11,000 years ago. This level is named the Gilbert level. The present level of the lake varies near 4,200 feet above sea level. Shorelines other than those named above are visible in different locations around the Bonneville Basin, but the highest level is the Bonneville level and the most pronounced shoreline level is usually the Provo level. These elevations given are the original elevations at the time they were formed. When lake water was lost, the crust was relieved of the weight and rebounded by as much as 200 feet in the center of the lake where the water was 1,000 feet deep. The present elevations of the lake levels are somewhat higher due to this rebound.

Older episodes of glaciation are expressed in a more restricted and subtle variety of features. Ordovician glaciation from 505 to 438 million years ago occurred when Gondwana was near the South Pole with late Ordovician massive glaciers, but alas, no Ordovician rocks are present in the Wasatch to record this event. While no glacial deposits are present in the Wasatch of either Mississippian or Pennsylvanian age, the rocks show cyclic sedimentation typical of rising and falling sea level as glaciers advanced and retreated. Extensive glaciation of Gondwana began in latest Mississippian with a three-fold increase in magnitude of sea level fluctuations driven by increase in ice volume. The alternating sandstone and limestone beds of the Humbug Formation record the sea level fluctuations. Mississippian to Permian glaciation began in latest Devonian as Gondwana moved across the South Pole. The Pennsylvanian portion of this episode is indicated by cyclical deposition of sediments in the Oquirrh Basin and its margins in the Wasatch Mountains and in the Paradox Basin of southeastern Utah. Precambrian glaciation is expressed as grooved and glacially scoured bedrock and overlying conglomeratic glacial debris with dropstones (ice-rafted boulders) in the Mineral Fork Tillite in the Wasatch Mountains.

Growth of continental ice sheets also caused rapid sea level changes in Early Cretaceous time. But these sea level changes cannot be recognized in the Wasatch Mountains because there are no marine deposits of Cretaceous age.

The last glaciation is the most prominent in the Wasatch Mountains and played a significant role in determining the present topography and the nature of the deposits that fill the upper portions of the three canyons and their tributaries.

Present-day Topography

The streams that drain the Wasatch flow from near the crest of the range westward into Salt Lake Valley and eastward into Heber Valley and the Provo and Weber River systems. These streams follow the slope of the mountains, and erosion of their beds keeps pace with the continuing uplift of the mountains along the active Wasatch fault. Only three streams completely penetrate the Wasatch: the Bear River, the Provo River, and the Weber River. These rivers originate very near each other in the high Uinta Mountains to the east. All three of these rivers drain ultimately into Great Salt Lake.

Long before the present Wasatch Mountain uplift began, an impressive mountain range was formed to the west. During the Late Cretaceous to Early Tertiary, the area of the Great Basin was higher than either of the areas represented by the Wasatch Mountains or the Uinta Basin, but possibly not higher than the Uinta Mountains. Streams draining this ancient mountain range flowed eastward to an ocean whose presence is now indicated only by the sediment that accumulated on its bottom. The ocean has long since disappeared. Then later the rivers flowed into lakes that had replaced the ocean. Beginning in the Oligocene, 38 to 24.6 million years ago, the western United States was uplifted as the mountain range in western Utah collapsed. At the time of inception of uplift on the present Wasatch Mountains, stream channels were in place that could be followed eastward across the Wasatch Mountains. When it came time for waters to flow westward from the Uinta Mountains rather than eastward, stream channels through the Wasatch may have been present for the Weber and Provo Rivers. The rivers were then superimposed across the rising Wasatch Mountains. The direction of flow of the rivers has thus reversed from former eastward flowing to present westerly flow.

The third stream to get through the Wasatch Mountains actually flows around the Wasatch Mountains rather than penetrating them. The Bear River flows northward through parts of Wyoming and Idaho, then is diverted southward toward the Great Salt Lake near Soda Springs, Idaho. The diversion is caused by basalt flows. Otherwise the Bear River would flow into the Snake River drainage.

The present-day topography of the Wasatch was produced by the uplift of the footwall of the Wasatch fault and the wearing away of the footwall by streams and glaciers. The glaciers occupied mountain valleys that had been carved by streams. Streams have in turn modified the glacial valleys. Even the broadest, u-shaped canyon of Little Cottonwood has its stream-cut channel in the middle.

Many of the high peaks are composed of rocks that are resistant to erosion. Scotts Peak is composed of Thaynes Formation that has been hardened by metamorphism near a large igneous intrusion. Mount Raymond is composed of Weber Quartzite. Twin Peaks, Superior Peak, and Mount Olympus are composed of quartzite of the Big Cottonwood Formation. Mount Majestic, Wolverine, Millicent, and Lone Peak are composed of igneous intrusive rocks.

The Trails

This book covers numerous trails in Mill Creek, Big Cottonwood, and Little Cottonwood Canyons, as well as Mount Olympus, Neffs Canyon, and Bells Canyon trails. Hiking trailheads are shown in Map 1. Each trail or group of trails is illustrated with a geologic and shaded relief map. Geology for the geologic maps is modified from Crittenden (1965a–d) and Crittenden and others (1966). Geologic map patterns and symbols are shown in Figure 5. The reader is encouraged to consult published trail guides for more detailed information on trail characteristics—for example, *Hiking the Wasatch,* by John Veranth (Wasatch Mountain Club, 1988), and *Hiking the Wasatch: The Official Wasatch Mountain Club Trail Map for the Tri-Canyon Area* (Wasatch Mountain Club and the University of Utah Press, 1994).

Access to the canyons was provided in pioneer times by roads built to obtain timber and to operate the sawmills that were located in each of the canyons. Later access included mining roads. Charles Keller (2001) describes many of the trails that follow old mining roads, logging roads, and timber slides.

Mill Creek Canyon Trails

Mill Creek Canyon is a northeast-trending canyon extending from the Wasatch Mountain front to the mountain crest and divide. A paved, two-lane road leads up Mill Creek Canyon from Wasatch Boulevard at 3800 South. The sedimentary rocks here also trend northeast and dip down to the north (see Figure 2). The canyon remains in a limited sequence of sedimentary strata. The Mill Creek Canyon road enters the canyon in Weber Quartzite, and then closely follows the Park City Formation for

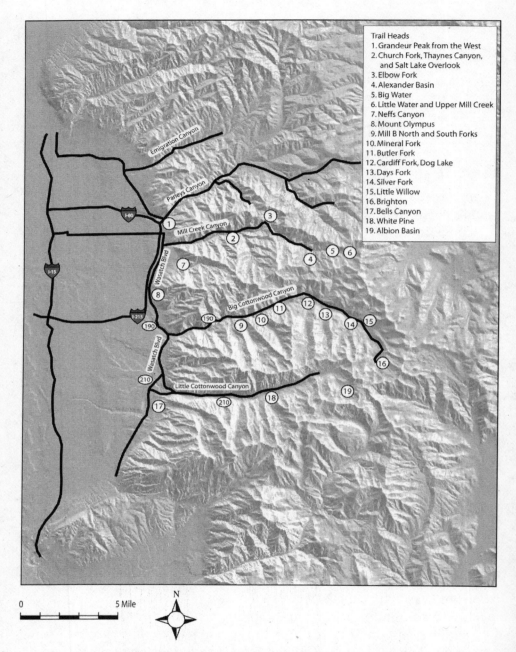

Map 1. *Shaded relief map of the Wasatch Mountains showing the main canyon roads and trailheads.*

most of its length. At Elbow Fork, the road enters the Woodside Shale and Thaynes Formation but quickly returns to the Park City Formation. The sedimentary rocks here are on the northern flank of the Big Cottonwood uplift with its crest to the south. Trails going north from the canyon rise upward through successively younger sedimentary layers, while trails to the south rise through successively older sedimentary layers.

The mouth of the canyon begins in the youthful sediments of Lake Bonneville, and the mountain front bedrock here is the Pennsylvanian-age Weber Quartzite that forms the prominent, stream-cut, v-shaped entry to Mill Creek Canyon. The canyon is steep sided and v-shaped for most of its length, but the upper portion of the canyon is more open and u-shaped, reflecting the glaciation that has shaped all of the upper Wasatch Mountains. Unusually fractured and broken rocks form pinnacles on the north wall of Mill Creek Canyon. The fractured nature and resulting erosion patterns of these rocks are due to thrust faulting.

Thaynes Formation shales and light gray, sandy limestone beds are very prominent on the north side of Mill Creek Canyon. Two thick limestone beds display the intricate folding of the Thaynes Formation that formed during Sevier-age (Cretaceous) mountain building.

Trailheads in Mill Creek Canyon are shown in Map 1.

Church Fork to Grandeur Peak

The trailhead at Church Fork is 3.2 miles into Mill Creek Canyon from Wasatch Boulevard. From the trailhead, the top of Grandeur Peak is 2.75 miles and 2,619 feet elevation gain. The geology of the trail is shown in Map 2. This can be a hot summertime hike as much of the trail lies on south-facing slopes in the oak and maple trees. The Church Fork trail proceeds northward from the Mill Creek Canyon road and rises through successively younger sedimentary strata. Prominent at the trail beginning is spring tufa, consisting of porous calcite (calcium carbonate) with roots and root tubes encased in calcite. The tufa formed by precipitation of calcite from saturated spring water flowing from underlying limestone formations. The trail from the end of the road in the picnic ground

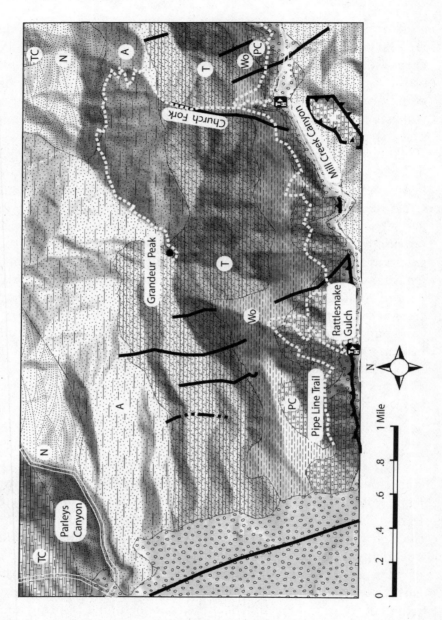

Map 2. *Shaded relief and geological map of the Church Fork and lower Pipeline trails. Geological map patterns are shown in Figure 5. Formation abbreviations are: PC = Park City Formation, Wo = Woodside Shale, T = Thaynes Formation, A = Ankareh Formation, N = Nugget Sandstone, TC = Twin Creek Limestone.*

follows a small normal fault in the Thaynes Formation. Several springs along the trail are probably controlled by this fault. The first 0.25 mile crosses prominent spring tufa. The trail follows the stream for about 0.5 mile of Thaynes Formation, then enters the Ankareh Formation shale and begins a series of switchbacks through oak and maple to a contact with the Nugget Sandstone. Nugget Sandstone near its lower contact with the Ankareh Formation is exposed from there to the ridge crest overlooking Salt Lake Valley and Parleys Canyon. The Gartra Grit portion of the Ankareh Formation forms prominent cliffs below the ridge crest. The trail then goes back into Ankareh Formation until the last short pitch to the summit of Grandeur Peak. The summit of Grandeur Peak is in the Thaynes Formation; the northeast ridge is in Ankareh Formation. The summit offers views of Salt Lake Valley, Great Salt Lake, Parleys Canyon, and the mountain-studded ridge dividing Mill Creek and Big Cottonwood Canyons. The southern view includes the prominent peaks of Mount Olympus, Mount Raymond, and Gobblers Knob, from west to east.

Grandeur Peak from the West

The trailhead is in a small parking lot with picnic tables at 2900 South Cascade Way. The distance to Grandeur Peak is 1.75 miles with a 3,100-foot elevation gain. A choice of routes is possible mostly in Thaynes Formation. Most routes up the west face of Grandeur Peak follow the ridgelines and remain in the Thaynes Formation to the summit of the peak. The trail routes are not marked in Map 2, but the ridges can easily be followed to Grandeur Peak. The route farthest north (closest to Parleys Canyon) begins in the deep red shales of the Ankareh Formation, but soon enters the Thaynes Formation, a brown-stained, fine-grained limey sandstone, olive green to dull red shale, and gray, fine-grained, fossil-bearing limestone. The Nugget Sandstone can be seen on the north side of Parleys Canyon. Prominent limestone beds in the Thaynes Formation are folded into a broad s-shaped fold that is visible from the valley. The fold formed during the Sevier mountain-building event. A few small north-south faults appear on the west face of Grandeur Peak.

Salt Lake Overlook

The beginning of the great Desolation trail is 3.4 miles from the canyon mouth. The overlook is 2 miles and 1,250 feet of elevation gain, while Thaynes junction is 3 miles and Porter Fork is 6.75 miles with 3,500 feet of elevation gain. The geology of the trail is shown in Map 3. This gentle trail has thirteen switchbacks to the overlook. It seems that you never get any higher. The trail is mostly in the Weber Quartzite, but after seven switchbacks (about 1 mile), one rock exposure is highly fractured and iron-stained limestone. This is an exposure of a hanging wall patch of thrust-faulted Park City Formation lying on the Weber Quartzite. The eighth switchback returns to the Weber Quartzite. The overlook, with a spectacular view of Salt Lake Valley, is in highly fractured Weber Quartzite. The view northward is Grandeur Peak, with the Thaynes Formation prominently displayed as limestone cliffs on its flanks. The ridge to the east of Grandeur Peak exposes Nugget Sandstone.

The trail continues an additional 1.25 miles to its connection with the Thaynes Canyon trail. This section of trail crosses the Round Valley Limestone and the Mount Raymond thrust fault that is covered by glacial moraine of upper Thaynes Canyon.

Thaynes Canyon

The Thaynes Canyon trailhead is in the same place as the Salt Lake Overlook trailhead and the geology is shown in Map 3. This is a steeper and more direct route to the upper portion of the divide between Mill Creek and Neffs Canyon. The trail follows Thaynes Canyon with few switchbacks unlike the Salt Lake Overlook trail. Charles Keller (2001) describes the log slide in Thaynes Canyon, and portions of the trail can be recognized as an old road. Because the sedimentary layers dip to the north, the trail proceeds into older rocks. The trail skirts a bit of Park City Formation at the bottom, and then goes into Thaynes Canyon in Weber Quartzite, through the Round Valley Limestone and the Doughnut Formation. A small spring seep marks the approximate location of the Mount Raymond thrust fault. After a short

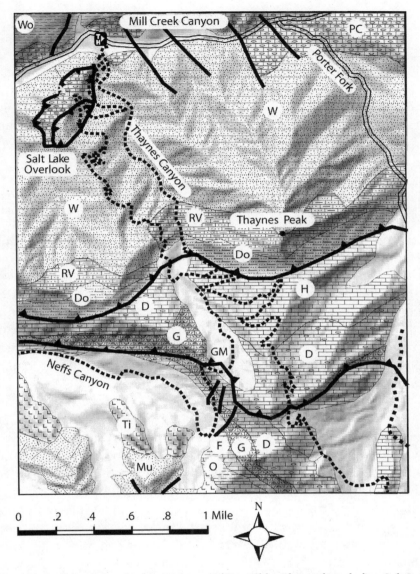

Map 3. *Shaded relief and geological map of Box Elder Flat trails including Salt Lake Overlook and Thaynes Canyon trails. Connecting trails to Neffs Canyon and Porter Fork are also shown. Geological map patterns are shown in Figure 5. Formation abbreviations are: PC = Park City Formation, Wo = Woodside Shale, W = Weber Quartzite, RV = Round Valley Limestone, Do = Doughnut Formation, H = Humbug Limestone, D = Deseret Limestone, G = Gardison Formation, F = Fitchville Formation, O = Ophir Shale, Ti = Tintic Quartzite, GM = glacial moraine.*

switchback through Humbug Formation, the Thaynes Canyon trail joins the trail from the Salt Lake Overlook (Desolation trail). One longer switchback in the Humbug Formation reaches the junction with the Neffs Canyon trail. The trail then follows glacial moraine in a high, glaciated bowl to the divide with Neffs Canyon in Deseret Limestone. The divide here is in Deseret Limestone, Gardison Limestone, and Fitchville Formation in the footwall of the lower strand of the Mount Raymond thrust fault.

Porter Fork Trail

The Porter Fork trail begins on a summer-home road, 4.4 miles up the canyon from Wasatch Boulevard. The road is 1.5 miles long and the pass to Big Cottonwood Canyon is 3.25 miles and 3,660 feet of elevation gain. Porter Fork is susceptible to avalanches from the slopes above. One man was killed here in 1852, three more in 1869, and another in 1889 (Keller 2001). The geology of the Porter Fork Trail is shown in Map 5. The trailhead is at the confluence of Mill Creek and Porter Fork in Park City Formation. The trail proceeds generally southeast, then more southerly with some twists and turns. Due to the north dip of sediments, the trail encounters older and older rocks as it ascends. The trail passes through the Park City Formation in less than 0.25 mile, then the trail is in the Weber Quartzite for all but the last 0.25 mile of paved road. The trail crosses 0.25 mile of Round Valley Limestone in the hanging wall of the upper strand of the Mount Raymond Thrust fault. The thrust is buried by glacial moraine on the trail and obscured by vegetation on the canyon walls. The canyon opens to broader glaciated topography. A few switchbacks take you into the upper bowl in glacial moraine underlain by footwall rocks that are Humbug Formation and Deseret Limestone. Then the trail crosses the lower strand of the thrust fault and goes back into the Humbug Formation in the footwall of the lower strand of the Mount Raymond thrust fault. The trail climbs steeply through forest and moraine to reach the divide in the Doughnut Formation.

The Scott prospect on the ridge between Porter Fork and Big Cottonwood Canyon may have produced some lead ore (James 1979).

The view southward from the divide includes Twin Peaks, Dromedary, and Sun Dial, all in Precambrian Big Cottonwood Formation. The trail skirts the flank of Mount Raymond in the Doughnut Formation and connects with the Mill B North Fork Trail. The summit of Mount Raymond exposes the resistant Weber Quartzite.

Pipeline Trail

The Pipeline Trail follows the grade of two separate pipelines (described in Parry 2004) that supplied water for generation of electric power at the upper Mill Creek power plant and a second power plant at the mouth of the canyon. The pipeline from Elbow Fork to Birch Hollow was a wire-wound, wood-stave pipe of redwood that was filled with water at Elbow Fork. The pipe transported the water to the penstock above the lower overflow parking lot. A few lilac bushes mark the location of the power plant. The second pipeline began at Birch Hollow and a larger, wire-wound, wood-stave pipeline of Douglas fir transported water to the mouth of the canyon, where a penstock with about 1,000 feet of vertical head was filled for power generation at the lower Mill Creek power plant. The overlook at the west end of this pipeline has a cast iron riveted Y. One arm of the Y connected with a surge pipe, and overflow from the surge pipe eroded a prominent gully down the west face of the ridge.

The geology of the Pipeline Trail is shown in Maps 2 and 4. There are four access points to the Pipeline Trail. The upper end of the trail is at Elbow Fork, 6.4 miles up the Mill Creek Canyon road. The Birch Hollow trailhead is 4.3 miles up the Mill Creek Canyon road. The Church Fork trailhead is just above the Church Fork picnic area, 3.2 miles up the Mill Creek Canyon road. The fourth access is at Rattlesnake Gulch about 1.5 miles up the Mill Creek Canyon road. From the intersection of the Rattlesnake Gulch trail and the Pipeline Trail, the trail continues west to an overlook of Salt Lake Valley. The Pipeline Trail is entirely within three sedimentary formations: the Thaynes Formation, Woodside Shale, and Park City Formation. From Birch Hollow to Elbow

Fork, the trail begins in the Park City Formation, crosses the brilliant red shales of the Woodside Shale, and ends at Elbow Fork in the Thaynes Formation. The trail east from Church Fork crosses Woodside Shale into the Park City Formation and remains in the upper Park City Formation to Birch Hollow. Occasionally some red shale can be seen from the overlying Woodside Shale. The trail west from Church Fork is almost entirely within the Park City Formation except for an occasional excursion northward in the gullies into Woodside Shale. At Rattlesnake Gulch, the effects of thrust faulting are evident. The Park City Formation is extensively broken and folded into a series of tight folds along the thrust fault exposed in Mill Creek Canyon.

Lime was produced from the Park City Formation in lime kilns at Rattlesnake Gulch and other locations in Mill Creek Canyon.

Bowman Fork

This trail begins at the Terraces picnic area 4.7 miles up Mill Creek Canyon and heads east up a narrow v-shaped canyon called Bowman Fork. The distance from the trailhead to Baker Pass is 3.75 miles with 3,080 feet in elevation gain. The geology of the trail is shown in Map 5. The trail begins in Park City Formation but soon crosses the contact into Weber Quartzite. After about 0.75 mile, the trail crosses a north-south normal fault and goes back into the Park City Formation, then the trail switchbacks southward up through Park City Formation to White Fir Pass at the contact of the Park City Formation and the Weber Quartzite. The trail then skirts around Yellow Jacket Gulch, which is in the Weber Quartzite. A steep traverse takes you through the Round Valley Limestone into the Doughnut Formation. The trail follows a small stream, crosses a ridge, and reaches Baker Spring near the Mount Raymond thrust fault in the Humbug Formation.

The Baker mine, also known as the Hayes mine, located on the northwest slope of Gobblers Knob, produced some rich gold ore with copper in 1904 (Calkins and Butler 1943).

The trail continues from Baker Spring across the western slope of Gobblers Knob in the Doughnut Formation and the Humbug Formation to reach Bakers Divide. Just before the divide the trail crosses the lower strand of the Mount Raymond thrust fault into the Park City Formation. Mount Raymond, visible from the divide, is composed of Weber Quartzite. On the way back down the trail—or behind you on the way up—you can see splendid views of Grandeur Peak and Mount Aire toward the northwest. These peaks are in the red Mesozoic age sediments. On the south flank of Grandeur Peak, the inclination of the Thaynes Formation is plainly visible.

Elbow Fork Trails

The Elbow Fork Trailhead is 6.1 miles up Mill Creek Canyon. The hike to Mount Aire is 1.8 miles with an elevation gain of 1,991 feet. Elbow Fork is so named because of the sharp bend to the right of the canyon road that follows the right hand or south fork of Mill Creek. The geology of the Elbow Fork trails is shown in Map 4. The trail goes up the left hand or north fork of Mill Creek named Elbow Fork. The trail to Mount Aire follows Elbow Fork in the Thaynes Formation for about 0.5 mile. Then on steep grades the trail enters the shale of the Ankareh Formation, which is very slick and muddy when wet. The Gartra Grit Member forms the ridge to the west. The Gartra Grit is a white-to-pink, coarse sandstone sandwiched between red shale above and below. The sandstone is an aquifer that may have conducted hydrocarbon reductants that bleached it white. The trail to Mount Aire proceeds to the right over more red shale and enters the oak trees. Shortly after the second switchback, the trail is in Nugget Sandstone to the summit. Through the branches of mountain mahogany and fir, Great Salt Lake appears. Stansbury Island in Great Salt Lake stands in the distance to the northwest with gray feet immersed in gray lake water. To the north, successive ridges finger out into the valley. Canyons separate the ridge crests in sequence to the north—first Parleys, then Emigration, City Creek, and beyond.

Elbow Fork to Lambs Canyon

The Lambs Canyon trail branches from Elbow Fork after about 0.2 miles in Thaynes Formation. The trail follows a tributary

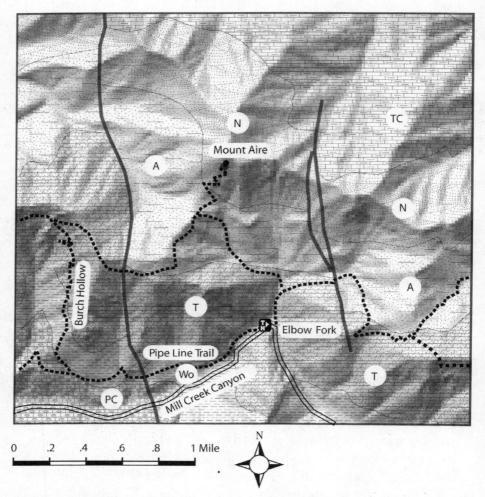

Map 4. *Shaded relief and geological map of Elbow Fork trails including Mount Aire, Lambs Canyon, and the Upper Pipeline Trails. Geological map patterns are shown in Figure 5. Formation abbreviations are: PC = Park City Formation, Wo = Woodside Shale, T = Thaynes Formation, A = Ankareh Formation, N = Nugget Sandstone, TC = Twin Creek Limestone.*

stream and crosses a small fault in 0.5 miles. The trail is in Ankareh Formation for about 0.5 miles then goes back into Thaynes Formation to the Lambs Canyon Divide. At the divide, a right-hand fork up the ridge is in Thaynes Formation and the trail down into Lambs Canyon is in the Ankareh Formation.

Alexander Basin

The Alexander Basin trailhead is 7.8 miles from the mouth of Mill Creek Canyon. The hike from the trailhead to Gobblers Knob is 2.6 miles with an elevation gain of 3,086 feet. The geology of the Alexander Basin Trail is shown in Map 5. According to Charles Keller (2001), the beginning of the trail follows straight up an old timber slide used to bring logs from the upper basin down to the sawmill near the Mill Creek stream. Alexander Basin is named for Alva Alexander who operated a shingle mill here. The nearest bedrock at the trailhead is Park City Formation both to the west and to the east of the trailhead. Alexander Basin is a glacial hanging valley perched above the v-shaped, stream-cut valley of Mill Creek. Glaciation did not affect Mill Creek for another mile upstream. Up the canyon road to the east is Thousand Springs issuing from the Park City Formation. The trail is exclusively in glacial moraine. Even the upper head wall is moraine until you get near the divide with Big Cottonwood Canyon. Some large boulders are probably from rock falls from the steep canyon walls. The trail follows the divide to the summit of Gobblers Knob. The Alexander Basin Trail within the hanging valley steepens, and then flattens again at each of a series of recessional moraines. Gobblers Knob is in the Pennsylvanian Round Valley Limestone and the Mississippian Doughnut Formations in thrust fault contact with the Weber Quartzite along the upper strand of the Mount Raymond thrust fault.

Big Water

The trailhead is at the lower Big Water parking lot at the confluence of Mill Creek and Soldier Fork 9.1 miles from the mouth of the canyon. The hike is 2.5 miles to the junction to Dog Lake. The geology is shown in Map 6. A switchback takes the trail to the west and then back eastward through thick fir trees. The trail then climbs gently to the intersection with Big Water gulch all in the Park City Formation. Soldier Fork, where the trail begins, is controlled by a fault related to the Mount Raymond thrust fault. The Silver Fork fault is at the confluence of Big Water with Mill

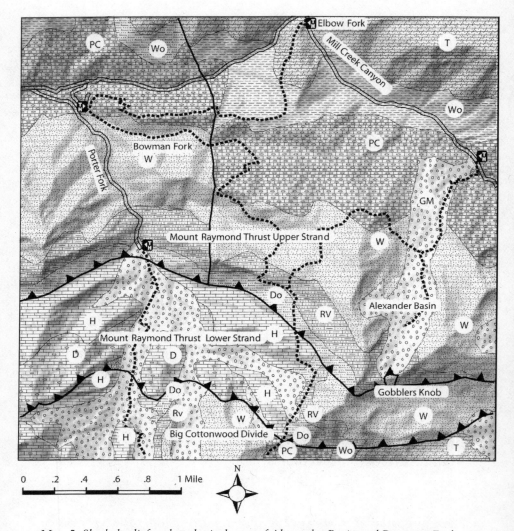

Map 5. *Shaded relief and geological map of Alexander Basin and Bowman Fork trails. Geological map patterns are shown in Figure 5. Formation abbreviations are: GM = glacial moraine, T = Thaynes Formation, Wo = Woodside shale, PC = Park City Formation, W = Weber Quartzite, RV = Round Valley Limestone, Do = Doughnut Formation, H = Humbug, D = Deseret.*

Creek. The trail follows this fault system to Dog Lake from the Park City Formation to the Ankareh Formation and back into the Park City Formation. Reynolds Peak is in the Nugget Sandstone, and the north flank of Reynolds Peak is in the Park City Formation in the upper plate of the lower strand of the Mount Raymond thrust fault. The Park City Formation has been thrust over the younger Nugget Sandstone.

Little Water

The Little Water trailhead is in the upper parking lot 9.1 miles from the mouth of Mill Creek Canyon. The trail is about 2 miles to the Dog Lake junction. The geology of the Little Water Trail is shown in Map 6. Here Mill Creek Canyon opens to a broad, glaciated bowl. The parking lot is on glacial moraine. The Little Water trail goes eastward parallel to but above Mill Creek for about 0.33 mile. It then crosses Little Water Creek and heads south up the creek. The trail soon crosses from moraine into the Park City Formation. After another 0.25 mile, the trail veers westward and crosses Little Water Creek again. A large spring at this location discharges from the Mount Raymond thrust fault where the Park City Formation and some Weber are thrust on top of the Ankareh and Thaynes Formations. After another 0.25 mile or so, the trail crosses a strand of the Silver Fork Fault where the Park City Formation is faulted against the Thaynes Formation. From here to Dog Lake the trail is back and forth in the Silver Fork Fault, the Mount Raymond thrust fault, and the Park City and Thaynes Formations.

Upper Mill Creek Canyon

The upper Mill Creek Canyon trail, a part of the Great Western Trail, branches from the Little Water Trail and the Big Water Trail about 1 mile from the trailhead. The geology of upper Mill Creek Canyon is shown in Map 6. The trail heads east through thick forest along the Mount Raymond thrust fault. The trail passes from the footwall of the lower strand in the Ankareh Formation into the hanging wall in the Thaynes Formation and then into

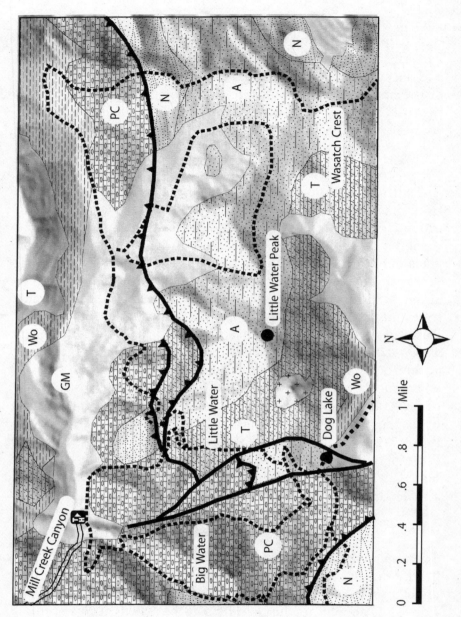

Map 6. *Shaded relief and geological map of Big Water, Little Water, and Upper Mill Creek Canyon trails. Geological map patterns are shown in Figure 5. Formation abbreviations are: N = Nugget Sandstone, A = Ankareh Formation, T = Thaynes Formation, Wo = Woodside Shale, PC = Park City Formation.*

upper-strand, hanging-wall Park City Formation. After about 1 mile the trail enters glacial moraine for a mile, then goes into the Park City Formation and Woodside Shale to the junction with the Wasatch Crest trail. The Wasatch Crest trail follows the ridgeline southward in the Park City Formation to the Mount Raymond thrust fault, crosses the thrust into footwall Nugget Sandstone, and then follows the Ankareh Formation to Desolation Lake.

Mount Olympus

The trailhead for the Mount Olympus hike is at approximately 5900 South on Wasatch Boulevard. The hike to the south summit of Mount Olympus is a strenuous 3 miles and 4,050 feet of elevation gain. The geology of Mount Olympus is shown in Map 7. The lower portion of the trail is in alluvium but soon reaches the purple quartzite of the Mutual Formation. Slickensides and breccias were formed on the Wasatch fault. Then a few switchbacks take you into the quartzites and shales of the Big Cottonwood Formation. The Mount Olympus south summit is in Big Cottonwood Formation quartzite. The south summit is separated from the north summit by more easily eroded shale of the Big Cottonwood Formation. The summit of Mount Olympus is approximately on the crest of a northeast-plunging anticline. Strata to the north of the summit are dipping steeply northward, while strata to the east are dipping steeply eastward. In fact, some of these eastern strata are overturned.

Neffs Canyon

The Neffs Canyon trail begins in Whites Park. Whites Park, located at the top of the neighborhood known as Olympus Cove, can be reached from the Mill Creek Canyon Road by turning right (south) on Parkview Drive. Follow Parkview Drive south, turn left on Park Terrace Drive, then turn right on White Way to the trailhead parking area. The distance to the meadow is 2.75 miles with an elevation gain of 2,450 feet. It is 3.5 miles to Thaynes Canyon pass and an elevation gain of 3,190 feet. Neffs Canyon geology is shown in Map 8.

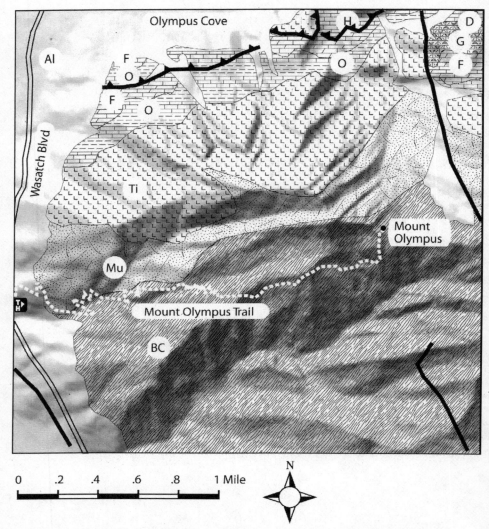

Map 7. *Shaded relief and geological map of the Mount Olympus Trail. Geological map patterns are shown in Figure 5. Formation abbreviations are: Al = alluvium, H = Humbug, D = Deseret, G = Gardison, F = Fitchville, O = Ophir Shale, Ti = Tintic Quartzite, Mu = Mutual Formation, BC = Big Cottonwood Formation.*

At the trailhead, the large, triangular ridge exposures of Tintic Quartzite on the north slope of Mount Olympus look like inclined flatirons. The trail follows an old road on the northern edge of an abandoned reservoir and several pipelines into the mouth of Neffs Canyon in alluvium. There are cast iron pipes, concrete pipes, and even one wooden pipe that brought water out of Neffs Canyon to irrigate farms below. The trail then crosses the lower strand of the Mount Raymond thrust fault in the Humbug Formation. The trail remains near the stream in alluvium, but the thrust faults and deformation in the Humbug Formation are visible on the north slope. The south slope is in the Tintic Quartzite. The trail then crosses a small normal fault into the Gardison Limestone, Deseret Limestone, and Fitchville Formation. The trail is in glacial moraine where the canyon opens into a broad, upper meadow. The trail to the Thaynes Canyon divide crosses the Gardison Limestone and Fitchville Formation, and the descent into Thaynes Canyon soon crosses the lower strand of the Mount Raymond thrust fault.

Big Cottonwood Canyon Trails

Big Cottonwood Canyon penetrates the Wasatch Mountains south of Mill Creek and south of Mount Olympus. The canyon road, Utah Highway 190, goes east from Wasatch Boulevard at about 7200 South. The canyon trends northeast through the northern flank of the Big Cottonwood uplift. Trailheads in Big Cottonwood Canyon are shown in Map 1.

Glaciers and streams shaped the topography of the canyon. The lower canyon exhibits a pronounced v shape typical of stream erosion, but the character of the canyon changes dramatically at Cardiff Fork where the broad u shape becomes prominent. At Cardiff Fork (Mill D South Fork), the Big Cottonwood Canyon glacier met the Cardiff Fork glacier. This intersection left behind some large moraines to the west and south of the intersection. Below this change in the canyon most of the tributary canyons are glaciated, but the Big Cottonwood stream has eroded the main canyon.

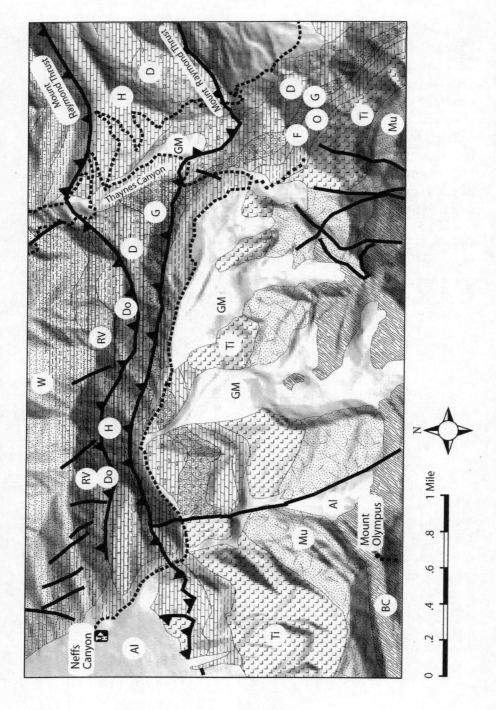

The mouth of the canyon begins in Lake Bonneville sediments. The lowest level is Provo age. Most of the Provo delta has been removed to supply sand and gravel, and a golf course has been constructed in the old gravel pit. A Bonneville-age delta lies above. Faulting of the lakebeds is prominent particularly to the south of the canyon mouth. The canyon above the mouth is cut into the Precambrian Big Cottonwood Formation. Just above Mineral Fork the purple Mutual Formation quartzite is exposed in the highway road cut. The highway crosses the salmon-colored Tintic Quartzite, then the Ophir Shale, followed by a thin sliver of Maxfield Limestone. The complete sequence of Paleozoic sediments is represented in road cuts. Fitchville Formation, Gardison Limestone, Deseret Limestone, Humbug Formation, and Doughnut Formation form Mississippian Limestone cliffs on the north side of the highway. The Pennsylvanian Round Valley Limestone and a very long road cut in Weber Quartzite occur up to Butler Fork. The brownish rocks in the Weber Quartzite are igneous dikes. They are altered and softer than the surrounding quartzite. Then, above Butler Fork, a long road cut exposes the Park City Formation that comes to an end at the Reynolds Gulch parking area. Part of the Park City Formation is very sooty and black. To the south is the broad u-shaped canyon of Cardiff Fork. Here the highway crosses the great Silver Fork fault into footwall Mississippian Gardison Limestone. Few outcrops are visible beneath the glacial moraine of the stream and roadbed. The highway continues on to the communities of Silver Fork and Brighton. Silver Fork is built on glacial moraine, but there is a small outcrop of Doughnut Formation just south of the highway east of Silver Fork. Between Silver Fork and Brighton the valley bottom is littered with large, erratic glacial boulders of igneous

---

Map 8. *Shaded relief and geological map of the Neffs Canyon trail. Geological map patterns are shown in Figure 5. Formation abbreviations are: Al = alluvium, GM = glacial moraine, W = Weber Quartzite, RV = Round Valley Limestone, Do = Doughnut Formation, H = Humbug, D = Deseret, G = Gardison, F = Fitchville, O = Ophir Shale, Ti = Tintic Quartzite, Mu = Mutual Formation, BC = Big Cottonwood Formation.*

rock from the igneous intrusives above. Brighton is also on glacial moraine, but a glacier carved the bowl into the igneous intrusives of the Alta and Clayton Peak stocks.

Trails to the south of the highway penetrate older rocks, and trails to the north of the highway penetrate successively younger rocks.

Approximately 20 percent of the land in Big Cottonwood Canyon is privately owned. Please respect private property rights when hiking the trails.

Mill B North Fork

The Mill B trailhead is 4.5 miles from the canyon mouth. A one-mile hike takes you to a splendid canyon overlook and 3.25 miles takes hikers to the Mill Creek Canyon divide on the Desolation trail. The geology of the trail is shown in Map 9.

The Mill B North Fork trail begins in the Big Cottonwood Formation. The trail switches back up a steep ridge of Big Cottonwood Canyon Formation. Alternating layers of white quartzite and thinner layers of shale were deposited in a great west-trending estuary. The source of these sediments was mostly the ancient continental basement rocks to the north in Wyoming, now exposed in the Teton and Wind River Ranges.

Several short switchbacks bring you to the top of a small ridge. The trail then drops down into the creek and follows the creek through the weakly metamorphosed quartzite and shale of the Big Cottonwood Formation. Then a couple of long switchbacks lead you through the Mineral Fork Tillite to the top of a quartzite ridge with a view to the south of the glaciated valley of Mill B South Fork. To see this view, leave the trail to the right and a short hike brings you to the ridge crest and the view. Along this trail, two panels of the Mineral Fork Tillite are exposed due to repetition by the Alta thrust fault about 800 feet above the trailhead. The Alta thrust fault is crossed three times on the trail from 800 to 1,400 feet above the trailhead. Here the Mineral Fork Tillite occupies two west-northwest-trending ancient glacial valleys eroded into the Big Cottonwood Formation. The southern valley is 3,000 feet

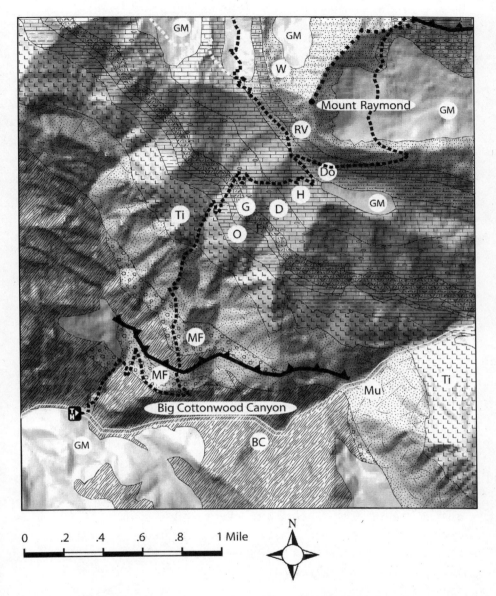

Map 9. *Shaded relief and geological map of the Mill B North Fork trails. Geological map patterns are shown in Figure 5. Formation abbreviations are: GM = glacial moraine, W = Weber Quartzite, RV = Round Valley Limestone, Do = Doughnut Formation, H = Humbug, D = Deseret, G = Gardison, F = Fitchville, O = Ophir Shale, Ti = Tintic Quartzite, Mu = Mutual Formation, MF = Mineral Fork Tillite, BC = Big Cottonwood Formation.*

deep and 4.5 miles wide, and the smaller northern valley is 650 to 1,400 feet deep and 1.25 miles wide. The contact with the overlying Mutual Formation (purple quartzite) is erosional with 50 feet of relief.

The trail then heads generally northward through younger sedimentary rock layers. The hanging wall of the thrust fault exposes the Big Cottonwood Formation and the Mineral Fork Tillite, then the trail crosses the purple Mutual Formation and descends into the gulch where the tan-to-salmon-pink Tintic Quartzite is exposed. At the head of the gulch, the trail sweeps upwards on switchbacks beginning with the Ophir Shale, then the trail crosses the Mississippian Fitchville, Gardison, and Deseret Limestones. The trail then goes around the nose in the Humbug Formation and follows the Doughnut Formation to the Mill Creek Divide. Mount Raymond to the east is in the Weber Quartzite.

Mill B South Fork—Lake Blanche

This trailhead is on the south side of the Big Cottonwood highway, 4.5 miles from the canyon mouth. Lake Blanche is 2.75 miles and 2,720 feet of elevation gain. The geology of the Lake Blanche trail is shown in Map 10.

The trail to Lake Blanche and its sister lakes Florence and Lillian lies in the broad, u-shaped and glaciated side canyon carved exclusively in the Precambrian Big Cottonwood Formation from its beginning near the s-curve on the highway to the lakes and beyond. Big Cottonwood Canyon at the trailhead has the characteristics of a stream-cut canyon. The glacial moraines here and at Broads Fork reached the level of Big Cottonwood Creek even though the main canyon is not glaciated. The trail is in Quaternary-age glacial moraine and younger talus from the steep cliffs on either side of the canyon. Glacial moraine is prominent along the trail. Quartzite forms both the east and west ridges for the first mile, as well as the outcrop near Blanche, Florence, and Lillian Lakes. Mineral Fork Tillite forms the East Ridge Crest after the first mile, so there are many boulders of this in the canyon. Lots of glacial grooved and polished rocks can be seen at

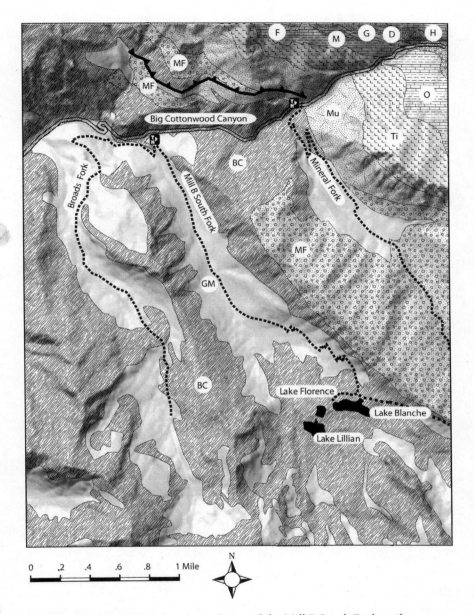

Map 10. *Shaded relief and geological map of the Mill B South Fork trails. Geological map patterns are shown in Figure 5. Formation abbreviations are: GM = glacial moraine, H = Humbug, D = Deseret, G = Gardison, F = Fitchville, O = Ophir Shale, Ti = Tintic Quartzite, Mu = Mutual Formation, MF = Mineral Fork Tillite, BC = Big Cottonwood Formation.*

the lake. One large boulder of igneous rock lies in the trail. Large glacial boulders along the trail consist of quartzite of the Big Cottonwood Formation and boulders of the Mineral Fork Tillite that have come from the ridge above to the east. Within 0.5 mile of the lakes, outcrops of the Big Cottonwood Canyon formation are prominent with glacial grooves, chatter marks, and other signs of the direction of ice movement down the canyon. The ice is too soft to have made such marks, but rocks embedded in the ice were dragged along the bedrock surface creating the marks. Sun Dial, Dromedary Peak, and, farther to the east, Superior Peak are prominent glacial horns along the ridge dividing the Mill B south fork drainage from Little Cottonwood Canyon. Near the lakes a series of small faults has broken the quartzites and shales.

This trail offers a fine comparison of the steep sided v-shaped, stream cut canyon of Big Cottonwood Creek and the broader, u-shaped, glaciated valley of the tributary Mill B Canyon.

The glacier that also left grooves, striations, and chatter marks on the polished rock surfaces excavated the lake basins holding Lake Blanche, Lake Florence, and Lake Lillian.

Broads Fork

The trailhead is in the same parking lot as the Mill B South Fork trail to Lake Blanche 4.5 miles from the mouth of the canyon. The hike is 2 miles to the end of the trail and 4.5 miles and 5,130 feet of elevation gain to the summit of Twin Peaks. The geology of Broads Fork is shown in Map 10.

Big Cottonwood Canyon is not glaciated for at least another 4 miles up canyon from Broads Fork. The Broads fork trail begins in glacial moraine from the Broads Fork glacier. Here, the trail is in a log slide used for bringing logs to the canyon bottom sawmills (Keller 2001). The trail remains mostly in moraine covering the Big Cottonwood Formation beneath it. The Big Cottonwood Formation is exposed here and there through the moraines. A few small faults cut the Big Cottonwood Formation. The hike is in Big Cottonwood formation all the way to the Twin Peaks summit.

Mineral Fork

The trailhead is just above the s curve of the highway and the trailhead to Lake Blanche, 6.1 miles from the mouth of the canyon. The trail is a mining road leading to the Wasatch (Silver King) mine, in 3 miles and 1,940 feet of elevation gain, and to the Regulator Johnson mine, in 4.5 miles and 3,510 feet of elevation gain. The geology of Mineral Fork is shown in Map 10.

Mineral Fork is in a glacial hanging valley perched above the stream-cut valley of Big Cottonwood Creek. The Mineral Fork moraines begin several hundred feet above the bottom of Big Cottonwood Canyon. Big Cottonwood Canyon is not glaciated until Cardiff Fork and Reynolds Flat are reached. The trail entrance to Mineral Fork is a switchback road that negotiates the steep hanging valley entrance. The hanging valley glacier of Mineral Fork likely produced much stream load here contributing to the steep reach of the stream down to Mill B. The entire canyon of Mineral Fork is cut into rocks of Precambrian age. The purple quartzite of the Mutual Formation is across Big Cottonwood canyon from the trailhead. This distinctively colored rock was deposited in a blanket controlled by sea-level rise at the end of the Precambrian "Snowball" earth glaciation. The trail switches westward into the older Big Cottonwood Canyon Formation that was deposited in a fault-bounded basin that extended from the open ocean to the west all the way eastward to the eastern end of the Uinta Mountains. As the road switches back to the northeast, the beginning of a large, northwest-trending glacial till is encountered. The Mineral Fork Tillite thickens and broadens in distribution all the way up the fork to the divide with Mill B South Fork. This is the type locality of the Mineral Fork Tillite. The tillite consists of a fine, dark matrix that encloses larger boulders of quartzite, occasional limestone, and rare igneous rocks.

The Wasatch Mine tunnel is farther up the canyon at an elevation of 8,680 feet (also called the Wasatch Gold Lower Tunnel or Silver King mine). The tunnel goes into the mountainside about 4,000 feet to intersect the northeast-striking, pyritic Silver King fissure that was also prospected in Mill D South Fork. A 2,000-

foot-long drift follows the fissure. The mine produced only 200 to 300 tons of gold, silver, lead, and copper ore, with the earliest production in the 1920s and the latest in 1952. The Silver King fissure contains lots of pyrite seen on the dump. The tunnel and drift opened the pyrite to oxidation and is the source of the iron coatings on stream gravels from the tunnel mouth downward.

The Regulator Johnson Mine, named for J. "Regulator" Johnson, is near the divide at 10,230 feet elevation. The Regulator Johnson mine is also on the Silver King fissure and shipped $280,000 in ore prior to 1914. An occasional fragment of the vein that was the Silver King fissure can be found in the piles of mine waste. The vein fragments contain brassy pyrite and quartz.

The view down Mineral Fork from the Regulator Johnson Mine shows the distinctive broad U shape of a glaciated hanging valley. Mineral Fork was occupied by glacial ice at a time when Big Cottonwood Canyon below was a vigorous mountain stream. A short, steep scramble brings you to the divide with Mill B South Fork, and a trail leads down to Lake Blanche.

Mill D South Fork (Cardiff Fork)

The trailhead is a mining road 9.3 miles from the mouth of the canyon. The trail to Doughnut Falls is 0.75 miles, the Cardiff Mine is 2.5 miles and 1,660 feet of elevation gain, and Cardiff Pass is 3.25 miles and 2,720 feet of elevation gain. The geology of Cardiff Fork is shown in Map 11.

There is something for everyone in this fork of the canyon. Cardiff Fork contains glacial deposits, mines, and faults. Cardiff Fork is the first glaciated tributary where the glacier reached the main Big Cottonwood glacier. The confluence of the two glaciers may have produced a glacier jam and prevented Big Cottonwood glacier from proceeding farther down canyon. The first mile of the road is in glacial moraine. Then the canyon and the trail follow the Superior Normal Fault on the right, or west, side.

The Price Tunnels are about 2.25 miles from the Canyon road intersection with Utah Highway 190 or 1.5 miles from the Doughnut Falls parking area. There are three tunnels here: the

West Price, East Price, and South Price tunnels. Fred W. Price, a native of Cardiff, Wales, who found the Cardiff ore body, sold his holdings in this rich deposit and attempted a repeat discovery here in 1915. The lower tunnel near the creek bed was extended for more than 1,000 feet in Tintic Quartzite and Price found nothing of value.

Faulting is the theme of upper Cardiff Fork. The ridge on the east is known as Reed and Benson Ridge, and is probably one of the longest continuous exposures of Cambrian to Mississippian limestone in the entire Wasatch Mountains. The ridge is about 3 miles long. It begins on the north with a small fold in the Weber Quartzite, and ends at the south in Precambrian Mineral Fork Tillite. The west-facing east cliff exposes Deseret Limestone, Gardison Limestone, Humbug Formation, and Fitchville Formation of Mississippian age, and the Maxfield Limestone and Ophir Shale of Cambrian age. The Ophir Shale occurs at the bottom near the talus slopes. Two small strands of the Alta thrust fault system are exposed on the east cliff. Several white bands are visible; the lowest is the white band in the Maxfield, and the upper two are a repetition of the white marker band in the Fitchville Formation by the thrust fault. The normal Superior Fault zone consists of a couple of separate strands, and the older Alta thrust fault is well exposed. The Mississippian Deseret Limestone and Gardison Limestone are faulted against the Tintic Quartzite in a small, normal fault graben. At the head of the canyon, Mineral Fork Tillite is on top of Tintic Quartzite and the Tintic Quartzite is on top of the Mississippian Limestones in thrust-fault contact and in reverse order of their ages. Cardiac ridge to the west is in Mineral Fork Tillite and Tintic Quartzite. Kesler Peak, visible to the north on the west side of Cardiff Fork, is in Mississippian Gardison Limestone.

The main tunnel of the Cardiff Mine, called the 600 level, is at 9,080 feet elevation. The ore was actually discovered at the number 3, or 300 level, tunnel above the main tunnel. Ore was in the Cardiff fissure, a northeast fissure in the Tintic Quartzite. The large ore body was found in 1914 at the intersection of the Cardiff fissure and the Alta thrust fault. The ore body was 10

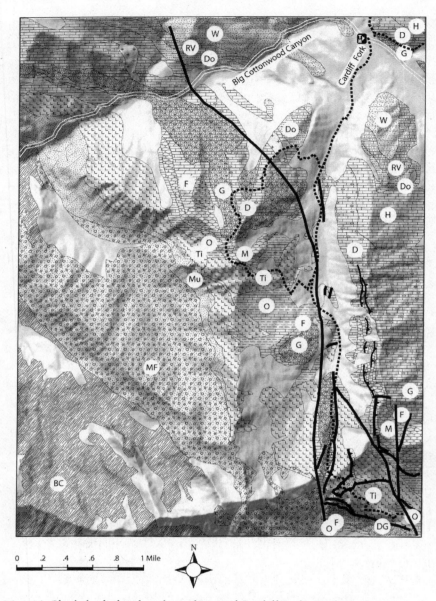

Map 11. *Shaded relief and geological map of Cardiff Fork. Geological map patterns are shown in Figure 5. Formation abbreviations are: GM = glacial moraine, W = Weber Quartzite, RV = Round Valley Limestone, Do = Doughnut Formation, H = Humbug, D = Deseret, G = Gardison, DG = Deseret and Gardison Formations undifferentiated, F = Fitchville, O = Ophir Shale, Ti = Tintic Quartzite, Mu = Mutual Formation, MF = Mineral Fork Tillite, BC = Big Cottonwood Formation.*

84

feet thick and extended for 300 feet down the plunge of the intersection. Other ore bodies were found on the thrust fault in Deseret Limestone. The mine produced 3 million ounces of silver along with copper and lead from ore that contained 20 ounces of silver per ton and at times was 20% lead. Nuggets of galena, tetrahedrite, enargite, and other sulfide minerals can still be found on the dump. The area was further prospected from the Wasatch Drain Tunnel that is near Snowbird on the Little Cottonwood side of the divide in the 1960s.

There is no water in the stream above the Price tunnels and Montreal spring on the Superior fault except in springtime. The Price tunnel is marked by conspicuous water flow. Surface water seeps into the Cardiff Mine workings and down to the Wasatch tunnel and provides water along with its dissolved load of arsenic, antimony, zinc, and other metals for Snowbird Resort.

Days Fork

The Days Fork trail begins in the Spruces Campground 10.1 miles from the mouth of the canyon. The trail on an old mining road is 2.75 miles to the Eclipse mine with an elevation gain of 2,250 feet. Silver Fork Pass is 3 miles and 2,570 feet elevation gain. The geology of Days Fork is shown in Map 12.

Springs near the mouth of Days Fork and the beginning of the trail flow from the Silver Fork Fault. The trail crosses from the Tintic Quartzite in the footwall side of the Silver Fork Fault into the Deseret Limestone on the hanging wall side; then, after about 0.25 mile of landslide debris, the trail is mostly in talus from the steep canyon walls. The east ridge is mostly Weber Quartzite.

Two miles up the canyon a number of mining prospects are in the Deseret Limestone, the Gardison Limestone, and the Humbug Formation.

The Eclipse shaft in the bowl at the end of the trail was sunk to a depth of about 500 feet beginning in 1881. The hoisting works burned down in either 1888 (Keller 2001) or 1893 (James 1979). The mine was worked with a steam hoist and steam air compressor driven by a wood-fired steam boiler. The remains of

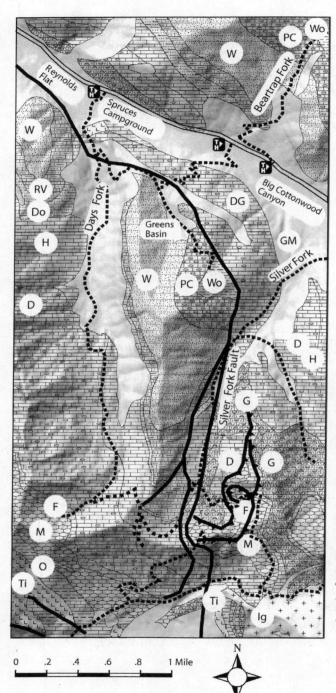

Map 12. *Shaded relief and geological map of Days Fork. Geological map patterns are shown in Figure 5. Formation abbreviations are: GM = glacial moraine, Ig = Igneous intrusive rock, Wo = Woodside Shale, PC = Park City Formation, W = Weber Quartzite, RV = Round Valley Limestone, Do = Doughnut Formation, H = Humbug, D = Deseret, G = Gardison, DG = Deseret and Gardison Formations undifferentiated, F = Fitchville, M = Maxfield Formation, O = Ophir Shale, Ti = Tintic Quartzite.*

0   .2   .4   .6   .8   1 Mile

N

the hoist drum and boiler can be seen in the debris in the shaft. Many of the tree stumps are no doubt the result of wood cutting for the wood-burning, steam-generating plant. The mine consisted of four levels extending southeast into the ore-bearing rocks that had been so productive on the Alta side of the ridge in the Emma and Flagstaff mines. In fact, the Eclipse underground workings do connect with the Flagstaff incline from Little Cottonwood Canyon. The ore, such as it was, was in the Fitchville Formation that had been so productive at Emma and Flagstaff mines and in the Gardison Limestone. The Fitchville-Gardison contact is just to the west of the Eclipse shaft, and the contact dips to the east down into the mine.

Butler Fork

The Butler Fork trailhead is located on the north side of the Big Cottonwood Canyon Highway 8.5 miles from the mouth of the canyon. Dog Lake on the right, or east, fork of the trail is 2.25 miles and the Mill Creek Canyon divide at Baker Pass on the left fork is 3.5 miles. The geology of Butler Fork is shown in Map 13.

The trail begins near the eastern end of a large road cut in the Weber Quartzite and passes through the Park City Formation. Near the trail junction, the right fork climbs through the Woodside Shale, the Thaynes Formation, the Ankareh Formation, and the Nugget Sandstone. The trail then crosses the Mount Raymond thrust fault into the Park City Formation. Just before reaching Dog Lake, the trail crosses the Silver Fork fault.

The left fork remains in the Park City Formation from the trail junction to Circle All Peak, which is in the Park City Formation near the contact with the underlying Weber Quartzite. The trail northward from Circle All Peak goes from the Park City Formation into the Woodside Shale and joins the loop trail from Dog Lake. The trail then follows the Woodside Shale and the Park City Formation to the Mill Creek Divide. The Woodside Shale here shows a prominent drag fold in the footwall of the Mount Raymond thrust fault, and the Park City formation is broken and veined as a consequence of deformation on the thrust

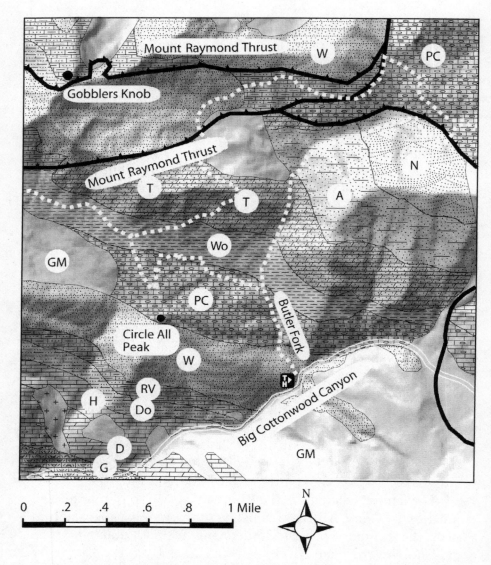

Map 13. *Shaded relief and geological map of Butler Fork trails. Geological map patterns are shown in Figure 5. Formation abbreviations are: GM = glacial moraine, N = Nugget Sandstone, A = Ankareh Formation, T = Thaynes Formation, Wo = Woodside Shale, PC = Park City Formation, W = Weber Quartzite, RV = Round Valley Limestone, Do = Doughnut Formation, H = Humbug, D = Deseret, G = Gardison.*

fault. Mount Raymond is in Weber Quartzite; Gobblers Knob is in Round Valley Limestone. Mill A basin is a prominent glaciated bowl.

Bear Trap Fork

The Bear Trap Fork trail begins on a road on private land 11.3 miles from the mouth of Big Cottonwood Canyon and just below the community of Silver Fork. The trail length is 2 miles to an overlook of Desolation Lake with a 2,240-foot elevation gain. A small peak to the east can be hiked with only a small increase in length and elevation. The trail geology is shown in Map 14.

Legend says that Bear Trap Fork is named for a bear trap that was seen by Sir Richard Francis Burton in his visit to the Big Cottonwood Canyon in September 1860 (Burton 1862). Burton wrote, "On our left, in a pretty grove of thin pines, stood a bear trap. It was a dwarf hut, with one or two doors, which fall when Cuffy tugs the bait from the figure of 4 in the centre."

The trail begins on a private road in a broad apron of glacial moraine from the Big Cottonwood glacier. At the end of the road where the canyon narrows, the trail follows a strip of alluvium between canyon sides of Weber Quartzite and then Park City Formation. The trail then goes into Woodside Shale that is not well exposed. The limestones and shales of the Thaynes Formation are obscured in the trees. Finally a steep hike on an indistinct trail through the upper bowl to the Desolation Lake overlook is in Ankareh Formation. The resistant band of Gartra Grit Member of the Ankareh Formation is exposed near the overlook. The view northeast from the small peak to the east of the overlook shows glaciated Dutch Draw and Red Pine Canyon in Nugget Sandstone and Twin Creek Limestone.

Willow Heights

The Willow Heights trail begins on the north side of the Big Cottonwood road just beyond Silver Fork, 12.1 miles from the mouth of the canyon. The geology of Willow Heights is shown in Map 14.

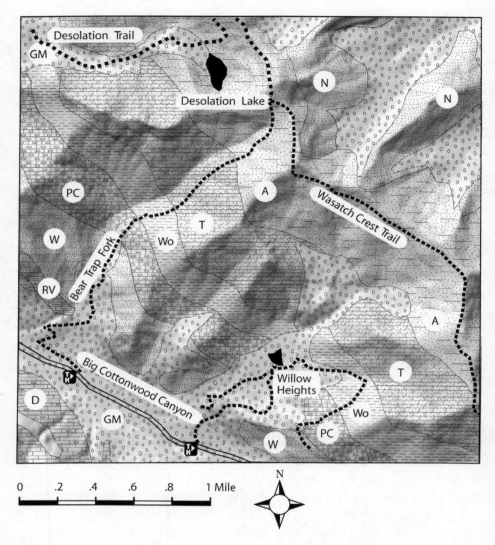

Map 14. *Shaded relief and geological map of Willow Heights, Bear Trap Fork, and Wasatch Crest trails. Geological map patterns are shown in Figure 5. Formation abbreviations are: GM = glacial moraine, N = Nugget Sandstone, A = Ankareh Formation, T = Thaynes Formation, Wo = Woodside Shale, PC = Park City Formation, W = Weber Quartzite, RV = Round Valley Limestone, D = Deseret.*

The trail follows Willow Creek to a small beaver pond, all in glacial moraine obscuring underlying Weber Quartzite, Park City Formation, and Woodside Shale. A continuation of the hike would include a hike cross-country up to Thaynes Formation and Ankareh Formation at the ridge crest divide with Snyderville Basin.

Dog Lake

The Dog Lake trailhead is at the Reynolds Flat parking area 9.6 miles from the canyon mouth. Dog Lake is 2 miles and 1,520 feet of elevation gain. Desolation Lake is 3.5 miles and 1,890 feet of elevation gain. The geology of the Dog Lake trails is shown in Map 15.

Glacial moraine deposits surround the trailhead at Reynolds flat. The trail soon crosses the Silver Fork fault, a normal fault with the footwall to the east. The trail crosses in a single step from the Permian Park City Formation across the fault into the Mississippian Deseret Limestone. The step from Park City Formation into Deseret Limestone represents a time span of nearly 100 million years. Just across the highway is a large pile of loose debris (moraine) deposited at the confluence of the Big Cottonwood and Cardiff Fork (Mill D South Fork) glaciers.

The trail climbs and curves around the hill and climbs through more Mississippian limestones. An overlook shows a view of a summer home area built on a small glacial moraine. The trail then crosses the Pennsylvanian Round Valley Limestone and then the trail is cut into the quartzite of the Pennsylvanian Weber Quartzite. Here the trail descends to the stream level. More Weber Quartzite is exposed, and then the trail goes back into the Park City Formation, here displaced nearly one mile northward along the Silver Fork Fault.

At the trail junction, the left fork goes to Dog Lake, while the right fork goes to Desolation Lake. The springs feeding the Mill D creek issue from the Park City Formation. The trail junction is at the top of the Park City formation in glacial moraine. This is a much smaller glacier than those on the south side of Big Cottonwood Canyon. The final mile to Dog Lake is in glacial

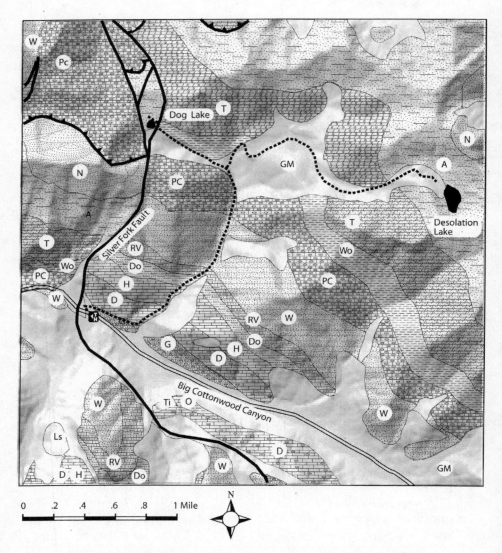

Map 15. *Shaded relief and geological map of Dog Lake and Desolation Lake trails. Geological map patterns are shown in Figure 5. Formation abbreviations are: GM = glacial moraine, Ls = landslide, N = Nugget Sandstone, A = Ankareh Formation, T = Thaynes Formation, Wo = Woodside Shale, PC = Park City Formation, W = Weber Quartzite, RV = Round Valley Limestone, Do = Dough-nut Formation, H = Humbug, D = Deseret, G = Gardison, M = Maxfield Formation, O = Ophir Shale, Ti = Tintic Quartzite.*

moraine and just skirts the bottom of the Triassic-age Woodside Shale. This is the Permian-Triassic boundary that represents a major extinction event. The ridge on the left of the trail is a glacial moraine.

Dog Lake is a small glacial lake that was left at the head of a small glacier and rests in a fault sliver of Ankareh Formation. Dog Lake is at a major fault intersection. The Silver Fork fault here has Ankareh Formation in the footwall and Park City Formation in the hanging wall, but the Park City Formation is repeated in the hanging wall by the lower strand of the Mount Raymond thrust fault.

Little Water Peak is northeast of Dog Lake. The peak is a small glacial horn with a diorite intrusive rock that is older than most of the igneous rocks in the Wasatch. The summit of Little Water Peak is in the Ankareh Formation that has been bleached, altered, and hardened by the igneous intrusive. To the southwest is Reynolds Peak in Jurassic Nugget Sandstone beneath the older Park City Formation on the Mount Raymond thrust fault.

Desolation Lake

The geology of the Desolation Lake Trail is shown in Maps 14 and 15. From the Dog Lake–Desolation Lake junction, the trail goes through glacial moraine, the Woodside Shale, the Thaynes Formation, and the Ankareh Formation. All of these rocks dip northeastward, so the trail goes up into younger and younger rocks. Desolation Lake is a glacial lake at the headwall of a small glacier. The lake is in Ankareh Formation. The band of white Gartra Grit Member of the Ankareh Formation crosses the mountain north and east of the lake. The Gartra is white sandstone in the midst of the red Ankareh Formation.

Silver Fork

The Silver Fork trail begins at the lower Solitude parking lot 12.7 miles up the canyon. The Alta Tunnel is 2.25 miles and 420 feet of elevation gain, the upper bowl and the Prince of Wales mine is 3 miles and 1,500 feet of elevation gain. The geology of Silver Fork is shown in Map 12.

Silver Fork joins Big Cottonwood Canyon in glacial till. Farther up the canyon the trail is largely upon till with bedrock of Deseret and Gardison Limestones on both sides of the canyon. After about one mile, the Silver Fork Fault system is encountered. The trail follows the fault, which dips gently west. The left, or east, fork of Silver Fork goes up to the Prince of Wales mine and the Grizzly (Alta) thrust fault. Here the Cambrian Maxfield Limestone is on top of the Deseret Limestone, Gardison Limestone, and Fitchville Formation.

The entire area to the southeast and west of the Alta Tunnel is heavily mineralized, and much silver and other metals were recovered. Groundwater was a problem for the early miners. Pumping the water out of mines with only vertical shaft access was expensive, so horizontal tunnels were driven at lower elevations to drain water from overlying strata and mines. The Wasatch Drain Tunnel in Little Cottonwood Canyon and the Alta Tunnel here in Silver Fork were constructed to dewater the mines. The deep Alta Tunnel, commenced in 1911, was driven 3,800 feet to drain the mines at Alta. A related objective was to prospect for new ore deposits. Only small deposits of lead, silver, and copper were found. The tunnel now serves as an important source of water today. The water flowing from the tunnel is probably more valuable than any metals that were ever recovered.

The trail leaves glacial moraine above the Alta tunnel and enters the Deseret Limestone, Gardison Limestone, and Fitchville Formation. The carbonates have been bleached and mineralized from interaction with heat and water from the igneous magma of the Alta Stock to the south. The contact of the igneous rock with sediments is gently inclined northward beneath this area. Fluid movement stimulated by igneous heat followed flow paths through the thrust faults and northeast-trending normal faults to produce the alteration.

In about 0.5 mile, the canyon forks into two steep forks, the left (east) fork heads up to the Prince of Wales mine following the Prince of Wales fissure with many mine workings. The right (south) fork heads into a glacial bowl with an ephemeral lake in the Humbug Formation. The Silver Fork and Snow faults are on the east side of the bowl.

The Prince of Wales shaft, at 9,875 feet elevation, is only part of extensive underground workings along the Honeycomb cliffs. The fissure was discovered in 1870 and was an early producer of silver in the district. The shaft is 930 feet deep following the Prince Fissure. In 1875, some 2,000 tons of ore containing 135 ounces of silver per ton and 35 percent lead were produced from the northeast fissure. This gave Silver Fork its name.

Snake Creek Pass

The trailhead is on the east side of the Brighton parking lot on glacial moraine. A 2-mile hike and 1,310 feet of elevation gain reaches Snake Creek Pass overlooking the Heber Valley. Clayton Peak is 0.5 mile farther or a total of 2.5 miles and 1,960 feet of elevation gain. The geology of the Brighton trails is shown in Map 16.

This trail, also called the Brighton Lakes trail, is mostly in glacial moraine, but leaves the moraine and goes into the porphyritic phase of the Alta Stock. A right fork leads to Lake Mary and Lake Catherine. Near Dog Lake the left fork of the trail crosses into the older, more mafic Clayton Peak Stock and remains in this lithology to the pass and then on to the summit of Clayton Peak. Snake Creek Pass overlooks Heber Valley, one of the faulted back-valleys of the Wasatch.

Brighton Lakes Trail (Lake Mary Trail)

The Lake Mary trailhead is on the east side of the Brighton parking lot, and is the first segment of the Brighton Lakes Trail. An alternative trail to replace the lower section of the old Lake Mary trail is currently being investigated by the Forest Service and environmental groups. Lake Mary on the right fork of the trail is 1 mile and 760 feet of elevation gain. The geology of Lake Mary is shown in Map 16. This trail is in the porphyritic phase of the Alta Stock. The trail continues past Lake Mary to Lake

Figure 17. *Lake Catherine in the shadow of Pioneer Peak. Pioneer Peak is Deseret and Gardison Limestones altered by metamorphism at the contact with the igneous intrusive of the Alta Stock. Lake Catherine lies in a glacial basin at the contact between sedimentary and igneous rock.*

Martha and Lake Catherine. Lake Catherine, the third lake on the Brighton Lakes Trail, is on the margin or contact of igneous and sedimentary rocks. Here, the Deseret Limestone, Gardison Limestone, Fitchville Formation, and Maxfield Limestone have been metamorphosed producing a sugary or sanded texture. Sunset Peak overlooking Lake Catherine is in the Maxfield Formation and the summit is in the Fitchville Formation. The Maxfield Limestone and Fitchville Formations are in the upper plate of the Alta-Grizzly thrust fault; the Deseret Limestone, Gardison Limestone, and Humbug Formation are in the lower plate. Mount Wolverine, to the north of Catherine Pass, is in the Alta Stock granitic phase. The trail continues beyond Lake Catherine to Catherine Pass and Albion Basin.

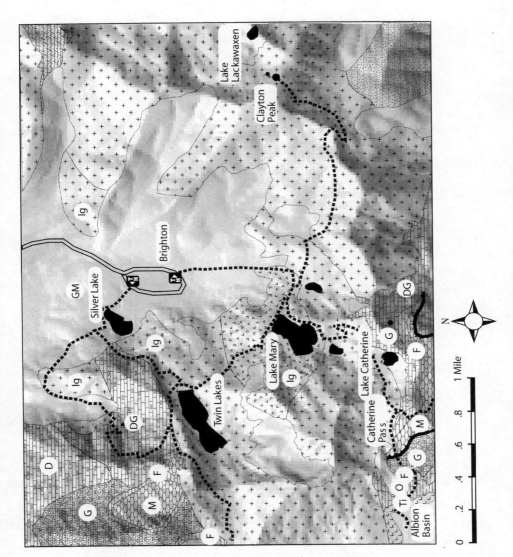

Map 16. *Shaded relief and geological map of Brighton trails. Geological map patterns are shown in Figure 5. Formation abbreviations are: GM = glacial moraine, Ig = igneous intrusive rocks, D = Deseret, G = Gardison, DG = Deseret and Gardison Formations undifferentiated, F = Fitchville, M = Maxfield Formation, O = Ophir Shale, Ti = Tintic Quartzite.*

Figure 18 *Twin Lakes and Mount Millicent.*
Millicent is igneous rock of the Alta intrusive.

Twin Lakes

The trailhead to Twin Lakes and Solitude Lake is at the parking lot for the Solitude Nordic Center. The Nordic Center is on the right side of the highway at the beginning of the one-way loop road around Brighton. Twin Lakes is 1.25 miles and 710 feet of elevation gain; Twin Lakes pass is 2.5 miles and 1,260 feet of elevation gain. The geology of the Twin Lakes Trail is shown in Map 16.

The trail proceeds from the parking lot over the boardwalk to Silver Lake, which lies in a cirque with headwall cliffs to the west composed of Deseret and Gardison Limestones, a wedge of sediment between Alta Stock on the south and a lobe of the Clayton Peak stock on the north. The trail climbs steeply in a cut across metamorphosed Deseret and Gardison Limestones with many different minerals. Then the trail passes into the Alta Stock and follows near the contact between igneous and metamorphosed sediments to Twin Lakes Pass. Fractures in the igneous rock have been mineralized with sulfides now oxidized to a brown limonitic coating. Mount Evergreen to the east is in Deseret and Gardison Limestones.

The view west from the pass includes Pfeifferhorn along the ridge dividing Little Cottonwood and American Fork Canyons. This peak is a glacial horn that exposes the Little Cottonwood intrusive rock. To the right (north) on the other side of Little Cottonwood Canyon, Mount Superior can be seen along with the exposures of Big Cottonwood Formation and the darker Mineral Fork Tillite.

Solitude Lake

Solitude Lake is 1.5 miles from Silver Lake and 300 feet higher. The right fork of the trail around Silver Lake curves around a ridge of Clayton Peak Stock as the trail enters Mill F South Fork. The geology of the Solitude Lake Trail is shown in Map 16.

The trail begins in glacial moraine, then passes into metamorphosed Deseret and Gardison Limestones. Solitude Lake lies in a glacial basin on the Solitude igneous dike. The Solitude tunnel, just west and above the lake, connects with mines on the Alta side in Grizzly Gulch. This tunnel is the lowest entry to the Michigan-Utah mines. Early miners hauled ore from the Solitude tunnel, but later mining operations hauled ore from the Alta side of the divide. The Kentucky-Utah tunnel is 870 below and 2,000 feet west of Solitude Lake. The lake level dropped several feet when the tunnel was completed in the 1940s with heavy inflows of water into the tunnel. The water in the Kentucky-Utah tunnel is probably more valuable today than any ore extracted from the tunnel in the past.

Bells Canyon

The Bells Canyon trailhead is at 10400 South Wasatch Boulevard. Parking is on the east side of Wasatch Boulevard. The geology of Bells Canyon is shown in Map 17.

Bells Canyon is famous throughout the West because it is a glaciated canyon and the glaciers reached the foot of the mountains and the elevation of Lake Bonneville. In only two places in the west, glaciers reached the level of the ancient lakes, here and Mono Lake in California. This occurrence permits a determination of the relative age of glacial advance and lake rise. Careful search and observation in the area shows that the lake deposits of sand and gravel overlie and are therefore younger than the coarse, unsorted glacial moraines. The hike is 3.75 miles and 4,100 feet elevation gain to the upper reservoir.

Significant geologic features of this trail include the Wasatch fault, Bells Canyon faulted moraines, and the Bells Canyon lower reservoir that lies on the Wasatch fault. Many large, expensive homes are built along this active segment of the Wasatch fault. The trail proceeds up a stairstep sequence of recessional moraines that record the recession of the Bells Canyon glacier. The first rock outcrops are of the Precambrian Big Cottonwood Formation that forms a prominent cliff on the north side of the canyon mouth. The remainder of the trail is in moraine and igneous rock of the Little Cottonwood igneous intrusive.

Little Cottonwood Canyon Trails

No better display of glacial features is available in the Wasatch than in Little Cottonwood Canyon. The u shape of the canyon begins right at the canyon mouth. A large, lateral moraine forms the southern wall of the canyon mouth. The Bells Canyon moraine is above and south of the Little Cottonwood moraine. Lateral moraines line the canyon from the canyon mouth and along the road cuts leading up the canyon. Tributary hanging valleys with their own sets of glacial features are carved into the Little Cottonwood quartz monzonite (granitic rock). Glacial horns are prominent on the high ridges up to the Albion Basin bowl with the

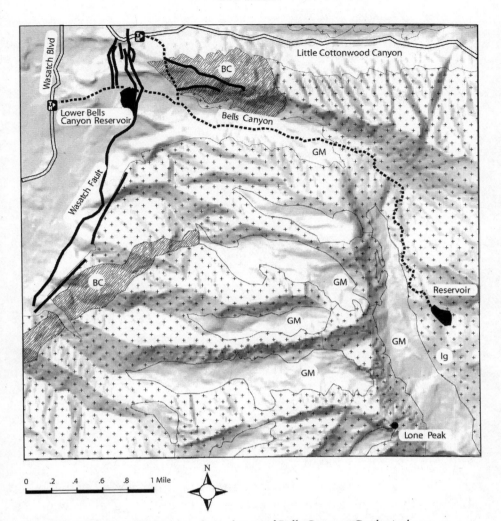

Map 17. *Shaded relief and geological map of Bells Canyon. Geological map patterns are shown in Figure 5. Formation abbreviations are: GM = glacial moraine, Ig = igneous intrusive of the Little Cottonwood stock, BC = Big Cottonwood Formation.*

Figure 19. *Bells Canyon waterfall.*

Devils Castle headwall. Aprons of moraine covered with younger talus merge with stream alluvium in the canyon bottom. Trailheads in Little Cottonwood Canyon are shown in Map 1.

Almost 20 percent of the land in Little Cottonwood Canyon is privately owned. Please respect private property rights when hiking these trails.

White Pine Canyon

The White Pine Canyon trailhead is in the broad glacial trough of Little Cottonwood Canyon 5.5 miles from the canyon mouth. White Pine Lake is 4.5 miles and 2,460 feet of elevation gain. The geology of White Pine Canyon is shown in Map 18.

According to Charles Keller (2001), Red Pine and White Pine Canyons were the locations of the earliest and most often used

Figure 20. *View of Big Cottonwood Canyon Twin Peaks from the south looking north. The peaks are Big Cottonwood Formation and the lighter rock beneath is the intrusive igneous rock of the Little Cottonwood stock.*

timber slides. Logging involved Engleman Spruce, called White Pine by the early pioneers, and Douglas fir, called Red Pine by the earlier pioneers.

Just below the trailhead, a large pile of angular boulders forms a transverse barrier across the canyon. A prehistoric rockslide from the north wall of the canyon formed this barrier. It is not a recessional glacial moraine. A moraine would likely contain fine glacial debris as well as the large boulders found here. The barrier must have partially dammed the stream because a large flat area (Tanners Flat) is filled with fine stream sediment.

The trail follows an old road to the White Pine stream, and then switches back and forth into the flatter mouth of the hanging valley cut by the tributary glacier of White Pine Canyon. The mouth of the hanging valley offers a spectacular view to the north of the Big Cottonwood Twin Peaks and the ridge dividing Big and Little Cottonwood Canyons.

The canyon is carved into the Little Cottonwood igneous intrusive emplaced 30.5 million years ago. This is the largest of the igneous intrusives in the east-west trend known as the Wasatch Igneous Belt. The igneous rock was emplaced at great depth, more than 6 miles deep. Exposure at the surface here is due to upward displacement of the footwall block of the Wasatch Fault which has also rotated about 20° eastward so the emplacement depth is greater to the west. The original ground surface is represented by volcanic rocks east of the Jordanelle Reservoir and northward in a crescent-shaped area to Peoa. As you walk the road leading to White Pine Lake, observe the large crystals of orthoclase feldspar that formed as the igneous magma slowly cooled 6 miles beneath the surface of the earth.

A younger intrusive igneous rock was emplaced within the Little Cottonwood stock. This younger rock carried tungsten, molybdenum, iron, and sulfur. The iron and sulfur formed pyrite—the pyritic phase—and weathering has stained the rock fractures with iron oxide. A mining company explored this area for low-grade molybdenum in the 1960s, and their drilling roads and drill locations can be recognized in the canyon.

The mineral composition of the Little Cottonwood stock is: 25 percent plagioclase feldspar, 40 percent orthoclase feldspar, 25 percent quartz, 7 percent biotite, and 3 percent hornblende, with traces of apatite, zircon, magnetite, and sphene. The texture is coarsely crystalline (phaneritic) with some large feldspar crystals (phenocrysts) reaching sizes from 0.5 to 2 inches.

A complex sequence of events led to the emplacement of the igneous rock and its exposure today. The sequence began with the collision of North America with the Pacific Ocean that led to great folds, thrust faults, and east-west shortening and thickening of the crust. The thickened crust then began to extend and flatten, accompanied by upwelling of hot mantle rocks and intrusion of iron- and magnesium-rich rocks and heat into the lower crust. The ancient Proterozoic crust then melted, forming buoyant magma that was emplaced along the Wasatch Igneous Belt.

Exposure of the deep-seated rocks occurred because of the continued extension and normal faults that fragmented the

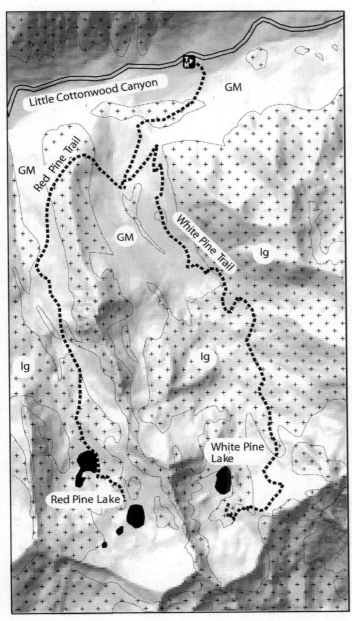

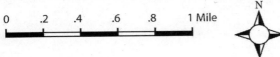

Map 18. *Shaded relief and geological map of the Red Pine and White Pine trails. Geological map patterns are shown in Figure 5. Formation abbreviations are: GM = glacial moraine, Ig = Igneous intrusive of the Little Cottonwood stock.*

thickened crust into north-south-trending blocks of valleys and mountains. Uplift and erosion of the mountain blocks exposed deeper, older rocks.

The uplift in the Wasatch Mountains is greatest along the continuation of the Uinta Mountains, here called the Big Cottonwood Uplift. The crest of this uplift is the most deeply eroded, and exposes the oldest rocks and the igneous bodies with the greatest emplacement depths.

The Little Cottonwood stock is the focus of a famous geological controversy. Clarence King, his field assistant S. F. Emmons, and his German petrographer Ferdinand Zirkel (King 1878) had identified the textures and structures of the rock, particularly the deformed margins, as typical of Precambrian age granite (note: radiometric dating methods were still almost a century in the future). Clarence King was the leader of the geological exploration of the Fortieth Parallel, and would become the first director of the United States Geological Survey. Because some 40,000 feet of sediment occurs at the contacts, King had proposed a 40,000-foot-high sea cliff against which the sediment was deposited. Later, inclusions of the sediment were found in the igneous rock establishing its relative age as younger than the sediment. James Boutwell in 1902 (published in 1912) demonstrated to Emmons that the Little Cottonwood stock was intrusive, and Emmons then published an acknowledgment in 1903 that the Fortieth Parallel Survey had made a mistake. Much later the absolute ages were measured by radiometric dating methods.

Red Pine Canyon

The Red Pine Canyon trail begins in the White Pine Canyon trailhead 5.5 miles from the canyon mouth. Lower Red Pine Lake is 3 miles and 1,940 feet of elevation gain; Maybird Lake is 3.75 miles and 1,946 feet elevation gain; and the Pfeifferhorn is 4.5 miles and 3,700 feet of elevation gain. The geology of Red Pine Canyon is shown in Map 18.

The Red Pine trail is much like the White Pine trail. The trail is entirely within either glacial moraine or igneous rocks of the

Little Cottonwood intrusive. A small mine prospect into the pyritic phase of the Little Cottonwood intrusive is about two miles up the trail, or one mile below Red Pine Lake. Red Pine and Upper Red Pine Lakes are tarns (glacial lakes), while Pfeifferhorn is a glacial horn. The hike up the trail presents magnificent views down Little Cottonwood Canyon showing the pronounced glaciated u shape and northward across the canyon showing the Little Cottonwood intrusive intruding the Big Cottonwood Formation.

Catherine Pass

The Catherine Pass trailhead is in a broad parking area at the top of the Sunnyside ski lift above Alta on the summer road to Albion Basin. Catherine Pass is approximately 1 mile and 800 feet of elevation gain. The geology of the Catherine Pass Trail is shown in Map 19.

The trail begins in igneous rock of the Alta intrusive. The trail follows the margin of the igneous intrusive, then enters Tintic Quartzite and Ophir Shale. Heat and fluids from the igneous intrusive have metamorphosed the sedimentary rocks. The trail then crosses into Deseret and Gardison Limestones that have also been bleached and metamorphosed to marble. The trail crosses a strand of the Alta-Grizzly thrust fault into Maxfield Limestone in the hanging wall that has also been transformed by heat and fluids. The trail leaves the Maxfield Limestone and enters igneous intrusive rock at the pass. Extensively metamorphosed limestones are exposed on the ridgeline from the pass to Sunset Peak. The cements holding the limestone crystals together have been removed and the rock is much like a sand of carbonate crystals. Lake Catherine is a glacial lake at the foot of Sunset and Pioneer Peaks and lies at the contact between igneous rock and the Deseret Limestone.

Grizzly Gulch to Twin Lakes Pass

The trailhead for this hike up an old mining road is located in the town of Alta near the Forest Service garage 8.3 miles from the canyon mouth. It is 1.75 miles and 1,353-foot elevation gain to

Figure 21. *Mine hoist and boiler at the Prince of Wales shaft at the head of the left fork of Silver Fork.*

Twin Lakes Pass, and 2.25 miles and 1,460 feet of elevation gain to the Prince of Wales mine.

This trail is as rich in mining history as in geological features. The trail traverses sedimentary rocks that are down-faulted along two of the major normal faults of the Wasatch: the Superior fault to the west with Superior Peak on its uplifted or footwall block, and the Silver Fork Fault to the east with Honeycomb Cliffs on its uplifted or footwall side. This block of sedimentary rocks is faulted down between two faults and is situated between two intrusive rocks of the Wasatch Igneous Belt, the Alta stock to the east and the Little Cottonwood stock to the west. The sedimentary rocks here have been extensively mineralized with rich silver deposits. The Flagstaff and Emma mines were the largest mines.

The trail follows a mining road eastward across Ophir Shale, Maxfield Limestone, and Fitchville Formation, and after about 0.75 miles crosses a small fault into the Deseret and Gardison Limestones. The Cambrian and Mississippian limestones are bleached by metamorphism. The ore was emplaced in the Fitchville

Formation with rich ore along northeast-trending ore shoots. The bleached limestones beside the trail are near the Emma mine.

According to Rossiter W. Raymond (1872), "The Emma Mine is one of the most remarkable deposits of argentiferous ore ever opened." J. B. Woodman discovered the rich silver ores while prospecting the North Star claim, one of the earliest mining claims at Alta. The Emma claim was located in December 1868. The ore did not outcrop; in fact, little evidence at the surface indicated the subsurface presence of the rich Emma ore. Some early locators claim 100-pound nuggets of galena were found in nearby ravines, an unusual claim because the ore body was completely oxidized and little sulfide remained. The ore mass was 60 feet by 40 feet and 130 feet high. Early production contained 100 to 216 ounces per ton of silver and was up to 45% lead. The rich ore was shipped by rail to New York, then by steamer to Liverpool, England, for smelting.

Owners forced maximum production between April and September 1871, and produced 5,638 tons of ore that was sold in Great Britain with a net profit of £158,268. The mine was sold to an English company capitalized at £1,000,000. The mine paid dividends of 1.5 percent per month until the end of 1872 when ore was terminated by a fault, causing a spectacular financial collapse of the English Company. This remarkable ore deposit was used as a basis of financial market rigging rather than as a mineral resource.

J. J. Beeson, a young geologist with a fresh university degree, studied the rocks and faults exposed on the Alta Ridge. In 1916 he discovered the faulted extension of the Emma ore that was displaced 250 feet down dip of the Montezuma fault. The limestones there all looked very much alike, but Beeson was able to work out the subtle characteristics of each limestone bed and from this, together with some scattered broken rocks, deduce the nature of the faulting that had so abruptly terminated the Emma ore body.

The trail above the Emma Mine workings is in Deseret and Gardison Limestones that are cut by the Silver Fork fault. The trail then enters igneous rock of the Alta stock just where the Alta

Figure 22. *Devils Castle at the head of Albion Basin. Fitchville Dolomite forms the bottom of the cliff and is overlain by Deseret and Gardison Limestones for most of the cliff face. The Ophir Shale is covered by talus and snow at the bottom.*

Grizzly thrust fault puts Maxfield Limestone on the Deseret and Gardison Limestones.

The Michigan-Utah Mines are on the footwall side of the Silver Fork fault. The Michigan-Utah mines include 65,000 feet of underground workings including the Solitude Tunnel that connects to the Cleves Tunnel in Grizzly Gulch. Ore here is also on northeast-trending fissures, the City Rocks fissure and the Grizzly-Lavinia fissure. Then hike into the Alta Stock intrusive to Twin Lakes Pass. The carbonate rocks near the contact with the Alta Stock have been metamorphosed through the action of heat, fluids from the intrusive rock, and fluids within the carbonate rocks to produce a diverse assemblage of metamorphic minerals depending on temperature and fluid composition.

Cecret Lake

The trailhead to Cecret Lake (the historical spelling is retained here) is in a parking lot near the Albion campground. Cecret Lake is 0.75 mile and 420 feet of elevation gain. The geology of

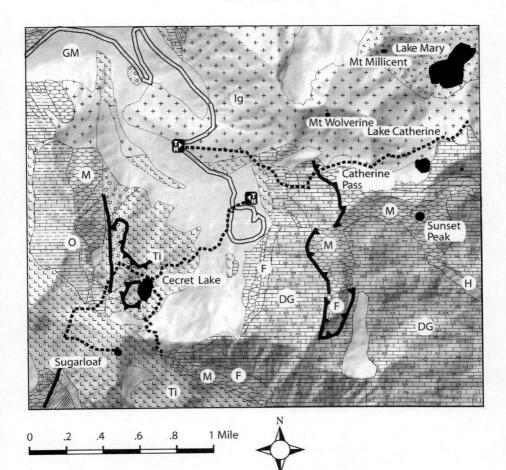

Map 19. *Shaded relief and geological map of the Albion Basin. Geological map patterns are shown in Figure 5. Formation abbreviations are: GM = glacial moraine, Ig = igneous intrusive of the Alta stock, H = Humbug Formation, DG = Deseret and Gardison Formations, F = Fitchville Formation, M = Maxfield Formation, O = Ophir Shale, Ti = Tintic Quartzite.*

the Cecret Lake trail is shown in Map 19. The Cecret Lake trail begins in the Albion parking lot in glacial moraine, but house-sized blocks of limestone are soon apparent. These blocks are part of a huge prehistoric rock fall from near the divide between Devils Castle and Sugar Loaf. The trail crosses to bedrock Tintic Quartzite with a few thin dikes of igneous rock. Small patches of Mineral Fork Tillite in thrust contact with underlying Tintic Quartzite are exposed near the lake. This is part of the Alta Grizzly thrust system that is so well displayed in the Hell Gate Cliffs area.

Deseret and Gardison Limestones are exposed on the Devils Castle Cliff. Near the base of the cliff, the prominent white band marks the Mississippian Fitchville Formation. These rocks, too, are in the footwall of the Alta Grizzly thrust fault that reappears farther northeast near Sunset Peak. If you hike on up to the Sugar Loaf–Devils Castle saddle, you cross a bit of the Maxfield Limestone and the Ophir Shale.

Cecret Lake is a glacial lake occupying the cirque of the Little Cottonwood glacier. Mount Baldy is in the Ophir Shale resting on the Tintic Quartzite in the upper plate of the Alta thrust. Beneath the thrust in the footwall are Deseret Limestone, Gardison Limestone, and Fitchville Formation that are visible on the west side of Mount Baldy.

Conclusion

Hiking through the geology of the Wasatch on these trails provides an experience in geology that is as varied and splendid as any region on earth. The effects of glaciation from 14,000 years ago to 700 million years ago are exposed here: rocks that accumulated in tropical seas on the continental shelf; rocks that accumulated in the largest field of sand dunes the world has ever known; rocks that accumulated in a great westward-flowing river system; rocks that formed from vigorous, eastward-flowing mountain streams; and sediments that accumulated in one of the largest and deepest glacial-age lakes in North America. The effects of numerous episodes of mountain building, including one episode that is in action today, are preserved here, waiting to be explored along the numerous Wasatch trails.

References

Bissell, Harold J., and Orlo E. Childs. 1958. The Weber Formation of Utah and Colorado. In *Symposium on Pennsylvanian Rocks of Colorado and Adjacent Areas*. Rocky Mountain Association of Geologists, Denver, Colorado.

Boutwell, J. M. 1912. *Geology and ore deposits of the Park City district, Utah*. U.S. Geological Survey Professional Paper 77.

Brandley, Richard T. 1990. Depositional history and paleogeography of the early to late Triassic "Ankareh Formation," Spanish Fork Canyon, Utah. BYU Geology Studies 36, 135–52.

Bryant, Bruce. 1990. Geological map of the Salt Lake 30'x 60' quadrangle, North-Central Utah, and Uinta County, Wyoming. Miscellaneous Investigations Series Map I-1944, U.S. Geological Survey.

———. 1988. Evolution and early Proterozoic history of the margin of the Archean continent in Utah. In *Metamorphism and crustal evolution of the Western United States,* vol. VII, ed. W. G. Ernst. Englewood Cliffs, NJ: Prentice Hall, 431–45.

Burton, Sir Richard Francis. 1862. *The City of Saints*. New York: Harpers. Reprint; Niwot: University Press of Colorado.

Calkins, F. C., and B. S. Butler. 1943. *Geology and ore deposits of the Cottonwood-American Fork area, Utah*. U.S. Geological Survey Professional Paper 201.

Christie-Blick, Nicholas. 1983. Glacial-marine and subglacial sedimentation upper Proterozoic Mineral Fork Formation, Utah. In *Glacial-Marine Sedimentation*, ed. Bruce F. Moline. New York: Plenum, 703–76.

Condie, Kent C., Dennis Lee, and G. Lang Farmer. 2001. Tectonic setting and provenance of the Neoproterozoic Uinta mountain and Big Cottonwood groups, northern Utah: constraints from geochemistry, Nd isotopes, and detrital modes. *Sedimentary Geology:* 141–42, 443–64.

Condie, K. C. 1997. *Plate tectonics and crustal evolution*. Boston: Oxford.

Constenius, K. N. 1996. Late Paleogene extensional collapse of the Cordilleran foreland fold and thrust belt. *Geological Society of America Bulletin* 108: 20–39.

Crittenden, M. D., Jr. 1965a. *Geology of the Draper Quadrangle, Utah*. U.S. Geological Survey Map GQ377.

———. 1965b. *Geology of the Dromedary Peak Quadrangle, Utah*. U.S. Geological Survey Map GQ378.

———. 1965c. *Geology of the Mount Aire Quadrangle, Utah*. U.S. Geological Survey Map GQ379.

———. 1965d. *Geology of the Sugar House Quadrangle, Utah*. U.S. Geological Survey Map GQ380.

Crittenden, M. D., Jr., F. D. Calkins, and B. M. Sharp. 1966. *Geologic map of the Park City West Quadrangle, Utah*. U.S. Geological Survey Map GQ 535.

Dalziel, I. W. D. 1995. The earth before Pangea. *Scientific American* (January): 58–63.

DeCourten, F. L. 2003. *The Broken Land*. Salt Lake City: University of Utah Press.

Ehlers, Todd A. 2001. Geothermics of exhumation and erosion in the Wasatch Mountains, Utah. PhD dissertation, University of Utah.

Emmons, S. F. 1903. The Little Cottonwood granite body of the Wasatch Mountains. *American Journal of Science*, 4th series, 16: 139–47.

Gilbert, G. K. 1890. *Lake Bonneville*. U.S. Geological Survey Monograph Number 1.

———. 1928. *Studies of Basin-Range Structure*. U.S. Geological Survey Professional Paper 153.

Gilluly, J. 1932. *Geology and ore deposits of the Stockton and Fairfield Quadrangles, Utah*. U.S. Geological Survey Professional Paper 173.

Gosse, J. C., J. Klein, E. B. Evenson, B. Lawn, and R. Middleton. 1995. Berrylium-10 dating of the duration and retreat of the last Pinedale glacial sequence. *Science* 268: 1329–33.

Granger, A. E., B. J. Sharp, M. D. Crittenden, and F. C. Calkins. 1952. Geology of the Wasatch Mountains east of Salt Lake City. In *Geology of the Central Wasatch Mountains, Utah, Guidebook to the Geology of Utah No. 8*, ed. R. E. Marsell. Utah Geological Society, 1–37.

Hintze, L. F. 1988. *Geologic history of Utah.* Provo: Department of Geology, Brigham Young University.

Hoffman, Paul F., and Daniel Schrag. 2002. The snowball earth hypothesis: testing the limits of global change. *Terra Nova* 14, 3: 129–55.

James, L. P. 1979. *Geology, ore deposits, and history of the Big Cottonwood Mining District, Salt Lake County, Utah.* Utah Geological and Mineral Survey Bulletin 114.

Keller, C. L. 2001. *The lady in the ore bucket.* Salt Lake City: University of Utah Press.

King, Clarence. 1878. *Report on the geological exploration of the Fortieth Parallel, vol. 1, Descriptive geology, vol. 2, Systematic geology.* Washington, DC: U.S. Government Printing Office.

Kirschvink, Joseph L., Eric J. Gaidos, L. Elizabeth Bertani, Nicholas J. Beukes, Jens Gutzmer, Linda N. Maepa, and Rachel E. Steinberger. 2000. *Paleoproterozoic snowball Earth: Extreme climatic and geochemical global change and its biological consequences.* Proceedings of the National Academy of Sciences 97, 4: 1400–05.

Knoll, A. H., Nicholas Blick, and S. M. Awramik. 1981. Stratigraphic and ecological implications of Late Precambrian microfossils from Utah. *American Journal of Science* 281: 247–63.

Parry, William T. 2004. The Mill Creek Canyon wooden pipelines. *Journal of the West* 43, 4(Autumn): 58–65.

Raymond, R. W. 1872. *Statistics of mines and mining in the States and Territories west of the Rocky Mountains for 1871.* U.S. Government Printing Office, 321–22.

Sonett, C. P., and Marjorie A. Chan. 1998. Neoproterozoic Earth-Moon dynamics, rework of the 900 Ma Big Cottonwood Canyon tidal laminae. *Geophysical Research Letters* 25, 4: 539–42.

Sonett, C. P., E. P. Kvale, A. Zakharian, Marjorie A. Chan, and T. M. Demki. 1996. Late proterozoic and Paleozoic tides, retreat of the moon and rotation of the Earth. *Science* 273: 100–104.

Stokes, W. L. 1986. *Geology of Utah.* Utah Geological and Mineral Survey.

Veranth, John. 1988. *Hiking the Wasatch.* The Wasatch Mountain Club.

Vogel, Thomas A., F. William Cambray, LeeAnn Fehr, Kurt N. Constenius, and the WIB research team. 1998. Petrochemistry and emplacement history of the Wasatch Igneous Belt. 35–46 in *Geology and ore deposits of the Oquirrh and Wasatch Mountains, Utah,* ed. David A. John and Geoffrey Ballantyne, vol. 29 of *Society of Economic Geologists Guidebook,* 2nd ed.

Wasatch Mountain Club. 1994. *Hiking the Wasatch: the official Wasatch Mountain Club trail map for the Tri-Canyon area.* Salt Lake City: University of Utah Press.

World Wide Web Resources

Brigham Young University Department of Geology

http://www.geology.byu.edu

links to virtual field trips, Utah resources

University of Utah Department of Geology and Geophysics

http://www.mines.utah.edu/~wmgg/

links to Utah geology and numerous geology web sites

U.S. Geological Survey Web Page

http://www.usgs.gov

Utah Geological Survey Web Page

http://geology.utah.gov

links to Utah geology and many other geologic links

Utah State University Department of Geology

http://www.usu.edu/geoldept/

links to local geology, lectures, field trips

Index

Note: page numbers printed in *italics* and followed by *f, m,* or *t* refer, respectively, to figures, maps, or tables.